History of Mechanism and Machine Science

Volume 24

Series Editor
Marco Ceccarelli, Cassino, Italy

For further volumes:
http://www.springer.com/series/7481

Aims and Scope of the Series

This book series aims to establish a well defined forum for Monographs and Proceedings on the History of Mechanism and Machine Science (MMS). The series publishes works that give an overview of the historical developments, from the earliest times up to and including the recent past, of MMS in all its technical aspects.

This technical approach is an essential characteristic of the series. By discussing technical details and formulations and even reformulating those in terms of modern formalisms the possibility is created not only to track the historical technical developments but also to use past experiences in technical teaching and research today. In order to do so, the emphasis must be on technical aspects rather than a purely historical focus, although the latter has its place too.

Furthermore, the series will consider the republication of out-of-print older works with English translation and comments.

The book series is intended to collect technical views on historical developments of the broad field of MMS in a unique frame that can be seen in its totality as an Encyclopaedia of the History of MMS but with the additional purpose of archiving and teaching the History of MMS. Therefore the book series is intended not only for researchers of the History of Engineering but also for professionals and students who are interested in obtaining a clear perspective of the past for their future technical works. The books will be written in general by engineers but not only for engineers.

Prospective authors and editors can contact the series editor, Professor M. Ceccarelli, about future publications within the series at:

Agamenon R. E. Oliveira

A History of the Work Concept

From Physics to Economics

Springer

Agamenon R. E. Oliveira
Applied Mechanics and Structures
Polytechnic School of Rio de Janeiro
Federal University of Rio de Janeiro
Rio de Janeiro
Brazil

ISSN 1875-3442 ISSN 1875-3426 (electronic)
ISBN 978-94-017-7918-0 ISBN 978-94-007-7705-7 (eBook)
DOI 10.1007/978-94-007-7705-7
Springer Dordrecht Heidelberg New York London

Softcover reprint of the hardcover 1st edition 2014

Printed on acid-free paper

Springer is part of Springer Science+Business Media (www.springer.com)

Preface

Distinguished scholar, Agamenon Rodrigues Oliveira, has combined his mechanical and historical skills in a very successful way in this brilliant book on the history of the work concept during pre and post-revolutionary France. His book is devoted to the history of classical mechanics, focusing on the role played by the concept of Work within the history of science. The result is based on a profound historical and epistemological approach which sheds new light on the foundations of science by examining this crucial physical concept and its framework based on the main concepts of space, time, mass and force as fundamental quantities for the development of the concept of work in the history of physics and mechanical sciences.

The book is organized into two main parts (*Conceptual Genesis* and *Instrumental Genesis*) and eight chapters. The book ends with References and an *Analytical Index*.

In Chaps. 1 and 2, Oliveira deals with the impact of the concept of work in society within the Marxist debate in order to "[…] remark the influence of Positivist philosophy in general historiography as well as in historiography of science […]" (p. 37). He then intelligently and gradually discusses the scientific concept of "Work" in the chapters that follow.

Chapter 3 is devoted to the concepts of energy and work in the history of mechanics. The author intelligently avoids banal modern assumptions (p. 71), focusing on the mathematical interpretation of the concepts up through the conservation principle in history. The history of virtual laws is taken into account with great lucidity (p. 78), thus distinguishing this work from other books on the history of mechanical science that have appeared since the *Principle of Virtual Work became* "[…] intensively used as an alternative method to the equations of equilibrium taken from Newton's laws for a given mechanical system" (*Ivi*).

Chapter 4 is dedicated to rational mechanics within a theoretical and empirical physical context, mainly from Lagrangian to Laplacian sciences where algebra applied to physics moves from mathematics applied to physics to physics–mathematics as a new discipline which emerged in the nineteenth century, particularly in France, Great Britain, Germany and Italy. This was a new methodological approach to solving problems that were physical in origin, where the quantities can be physical and mathematical at the same time (first innovation) and measurements are not however a priority or a prerogative (second innovation) to establishing a coherent and valid physical science. Therefore, the emergence of physics–mathematics (discipline) belonged to physics, not to an advanced use of mathematics to solve physical problems; physics changed its *face* since within this new physical mathematical discipline it had changed its foundations.

The second part of the book starts with the development of the concept of work in the history of mechanics in Lazare Carnot's memoires (1779, 1781). Since he aptly focuses on the fundamental importance and role played by Carnot in the development of the concept of work, I am not surprised that Oliveira profoundly observes that Carnot's mechanical essays constitute an "[…] important link between d'Alembert and Lagrange […]" (p. 127). I have also examined mechanics and complex French mechanical and thermodynamic sciences. As a historian, I believe that Oliveira's effort to profoundly examine the crucial role played by virtual issues in the work of Lazare Carnot (1753–1823) and other important French scholars will be greatly appreciated by historians of science and researchers in mechanics. In particular, Oliveira maintains that the foundations, impact and heritage of Lazare Carnot's mechanics (Chaps. 5 and 6) have resulted in the formulation of a law of virtual work which is generalized for dynamics: the mechanics of the impact of hard bodies, virtual velocity and (geometric) motion demonstrate how the *Organisateur de la victoire*, *l'hombre des machines en général*, succeeded in formulating a generalized virtual law to evaluate the velocities of a system of hard bodies of which the initial velocity is known. A special debate on the views of the scholars of *École polytechnique* is also included (Chap. 7).

Finally, in the conclusion, general comments are presented from historical, political and economic standpoints, where "Being more concerned with a history of applied mechanics where the principal characters are renowned figures whose names frequently appear in engineering manuals [the final part of the book] represent a modest contribution to the history of mechanical engineering. It should be highlighted that science history texts dealing with the more applied sciences are very rare. Normally they are histories of the basic disciplines, such as mathematics, physics, chemistry or biology." (p. 208).

Oliveira has written an interesting scientific and historical essay revealing an impressive knowledge of the vast amount of original and modern publications regarding the history and epistemology of science; he helps readers avoid losing track of the historical path of arguments dealt with by using reasoning, summaries and illustrations. From a methodological and historical point of view I am also particularly appreciative of the effort to produce a history of science based on standpoints in history which are interpreted epistemologically.

The book's composition makes for absorbing reading.

Agamenon Rodrigues Oliveira has produced an important study and a significant contribution of support for mechanics, engineers, physicists, historians and epistemologists of mechanical and mathematical physical sciences.

Please enjoy this stimulating read!

Lille, France Raffaele Pisano

Contents

Work Concept in the Machines Context

Labour is the source of any wealth, stated the economists. Hence, besides the nature it provides the materials that are converted in wealth. The labour, however, is much more than that. It is the basic and fundamental condition of whole human life. And in such intensity that we can assure that labour has created the man itself.

(Friedrich Engels—On the role of labour in the transformation of ape in man, A cultural brochure, Causa Operária Editions, 2011)

The machines can be defined, classified and studied in their evolution according to any criterion as follows: power, complexity, utilization of physical principles, etc. On the other hand, at the beginning, one is obliged to choose between two ways of different thinking. The first one is the engineer point of view, who regards the technology, mainly the internal relations which tend to define the machine with relation to itself, as a technical fact. The second is a social approach, which study the technology in its connections with the society and defines the machine in relation to human labour, and as a social device.

(Harry Braverman—Labour and Monopolist Capital—Rio, Zahar Editors, 1977)

Introduction

The inspiration for this book about the concept of work came mainly from the writings of François Vatin, in particular his book entitled *Work: Economics and Physics* 1780–1830. In this short book Vatin, in a clear and didactic way, explains some ideas that traditionally have been treated in different fields of knowledge. In addition he highlights some clues about issues that have been of particular concern to us during the last five years. Another interesting motivation came from a specific problem studied and developed by Vatin, the complex process of incorporation of the physical concept of work into economic thought. He attributes the accomplishment of this task mainly to Louis-Henry Navier (1785–1867), Gaspard-Gustave de Coriolis (1792–1843) and Jean-Victor Poncelet (1788–1867).

The original idea underlying our concept of work was progressively enriched until it became clear that we needed to develop our investigation in two ways that were different from Vatin's book. First of all it was our intention to go in some depth into the theoretical and historical aspects of the basic concepts necessary to understand the concept of work. Obviously we need to study the concepts of space, time, mass and force as a kind of decomposition of the concept of work in its "elements". The second fundamental difference from Vatin's socio-economic approach was to contextualize the evolution of that concept within the development of natural sciences; thus our book in fact falls within the literature of the History of Science.

The key-idea that will be developed in this text is that the work concept is of great importance in Physical Sciences but it occupies a fundamental position in the development of Machine Sciences. Initially, our study discusses the relations between the ideas of work and energy starting from the ancient framework within which simple machines were created and used. In this context, behavior of machines was more recently expressed by a statement of Lazare Carnot (1753–1823): *what is saved in velocity is lost in force and vice versa.*[1]

The specific contribution of our book is an attempt to show the evolution of the physical concept of work in its internal development starting from some basic and original ideas and finally demonstrating how the work concept is used in theories of machines. After decomposition of the concept and its historical development, an investigation of Lazare Carnot's mechanics will be performed where that concept appears in a key role. Using a different approach from Vatin's book, one which presents applied mechanics and its applications in industrial mechanics but emphasizing socio-economic aspects, our goal is to study the development of Machine Sciences related with the living forces principle or in a modern characterization by applying the work-kinetic energy theorem to machines. This fact can also be regarded as a fundamental step in the process of application of the Newtonian framework of machines. In addition, *ipso facto,* this operation has a very important meaning from an epistemological point of view.

It is worth emphasizing that the term *evolution*, when used, has the same meaning as that used by Darwin in his masterpiece. The path described by the work concept in our investigation starts from those above-mentioned ancient ideas, proceeds through a number of transformations until being formalized and in a final step being incorporated into economic thought.[2] In other words, the process of *evolution* does not mean a progressive or linear development passing from an inferior step to a superior one. There is no one teleological sense in its development. With respect to *the machines context* that appears in some sections of the book, it seemed from the beginning to be richer in potentialities from the History of Sciences point of view.

General Considerations on Labour

In general labour, as we know nowadays, fulfills three basic needs of man: surviving and fulfilling economic needs, cultural and artistic creation associated with psychological activities and finally cooperation that characterises social and

---

[1] See page 194, Chap. VI.

[2] If we need to study the concept of evolution from a broader point of view we can see the book of Peter J. Bowler (Bowler 1989). Bowler discusses several conceptions about the term evolution. It is clear that the Darwinian concept does not suppose a sense of finality; in other words, there is no teleology in Darwin's theory of evolution.

political relations. As result, in any way that it can be considered, the category of labour represents a vast and complex field of investigation (Vatin 1998).

If we look at the meaning of the word, labour involves a dialectic between two oppositions: enjoying and suffering. This can be better understood by the origin of the words used to designate labour. In Latin the word labor comes from *laborare* that means punishing, or walking while shouldering a heavy load. In French these ancient terms were replaced in the fifteenth century by *travail,* a word that designated a piece of wood used to tie down animals, or three poles used to torture the condemned.

On the other hand, if we look at Sanskrit root *rabh* from which the verbs *arbeiten* and *laborare* are derived, it is easy to see that in its origin labour had a positive meaning as actuating with energy. Yet, all the words that derive from the Latin term *opus* are identified with beauty and creation. Depending upon the social conditions where labour appears, one pole or its contrary is more accentuated.

In ancient societies where slavery was dominant, labour in mines, workshops, in repairs of roads and other services were performed by slaves. This kind of production became the more general form of production everywhere without presenting a uniform character. The development of trade relations, mainly in the Mediterranean area after some degree of its development, demanded a more sophisticated interchange of products, by putting the quality of products as a new exigency. This need transformed slave labour itself and accelerated a set of changes in this mode of production. Further development transmuted some slave labour to a less binding servant relationship (Bowler 1989).

At the time of barbarian invasions of the Roman Empire a new kind of labour known as servitude became dominant. It was a result of wars after which the conquerers inevitably imposed massive vassalage among the defeated. The first form of servitude appeared in Egypt. It became the dominant form of labour in agriculture.

Other forms such as leasehold and partnership and small property only appear in the middle ages. In classical antiquity, farm properties were characterized by great extension associated to systems of slavery and servitude. The exception was the presence of free workers.

Perry Anderson in his *Passages from antiquity to feudalism* has proposed a chronology of this mode of production in Western Europe: *feudalism in Western Europe arose in the tenth century, expanded during the eleventh century, reaching the maximum of its development at the end of the twelfth century. In the thirteenth century, European feudalism produced a kind of unified and developed civilization, which presented a great advancement with respect to communities in the middle ages* (Jaccard 1974; Barret 1975).

Anderson finalizes his book by summarizing the crisis and the fall of feudalism as follows: *the medieval world, then, has finished in a generalized crisis. Not only the cradles of feudalism in eastern society but the western territories, where it had propagated or where it couldn't develop itself were places of profound process of dissolution and socio-economic mutation at the beginning of the fifteenth century. In the early modern period, in which Constantinople's trenches fell fighting against Turkish cannons, its consequences to political order in Europe still stay hidden. The final step to a state system which appeared from this process is nowadays unexplored* (Anderson 1991).

From Manufacture to Machinism Expansion

With the advent of industrial capitalism, labour changed radically. The artisanal production which supported manufacturing grew vastly toward amplified industrial production. In England, manufacturing was dominant from the middle of the sixteenth century until the end of the eighteenth century. The most significant change in labour was its decomposition; the final product, the merchandise, was no longer obtained by the individual action of an independent worker but became the result of a collective action of a productive unit or workshop. In this case each individual performs a partial operation. This process created a division of labour. Another important characteristic of manufacturing became a low utilization of machines except in more simple processes with intensive human or animal strength.[3]

The division of labour introduced by manufacturing not only specialized workers but confined them in a unique place and created a social labour organization which changed production towards capital interests. The artisanal skills were the production basis and were founded in the hierarchies created by the productive system.[4] The weakness presented by this system, which was soon modified, was the absence of machines. With this characteristic, capital had no way to attract workers to partial tasks; production still worked with a narrow technical basis and it had to eliminate threats to discipline questions which can cause modifications in labour rhythm. Otherwise the utilization of force was required. These conditions show the necessity of developing and introducing machines in economic production in order to eliminate the limitations and weaknesses of manufacturing and its dependence on individual skills as a regulation of production principles.[5]

---

[3] Marx has characterized the manufacturing process as a form of labour based on cooperation: "The cooperation is founded on the division of labour and assumes its classic form in manufacture. It is dominant and assumes this characteristic in capitalist mode of production during the manufacture period which initiates in the middle of the sixteenth century until the last third of the eighteenth century". (Capital, Vol. I, Chap. XII, p. 386 (Marx 1968)).

[4] About the origins of manufacturing, Marx wrote: "Manufacture has its origin and starts from artisanal labour, following two ways. Firstly, it appears from the combination of independent crafts which lose its independence and became so specialized that constitutes only partial operations of the production process of one merchandise. The other has its origin in cooperation of craftsmen of a given craft, by decomposing the craft in its different particular operations, isolating and individualizing to transform each one dependent only of a special worker. Manufacture, sometimes introduces the division of labour in a given production process, sometimes improve it or combines previous different crafts. Anyhow, however, its starting point, its final result is the same: a mechanism of production whose organs are human being". (Capital, Vol. I, Chap. XII, p. 388 (Marx 1968)).

[5] About the limitations of this type of work, Marx said: "It is because craftsman professional skill is the objective of production process, the worker is obliged to perform a partial function and his labour-force became forever an organ of this partial function. Finally, manufacture division of labour is a kind of particular cooperation and some of their advantages are not due to this particular form but by the general nature of cooperation". (Capital, Vol. I, Chap. XII, p. 389 (Marx 1968)).

It is interesting to remark that at this time, the manufacturing period, modern science arose; it was Galileo Galilei's (1564–1642) time and Descartes' (1596–1650) mechanist philosophy. A new world vision appears in contradiction with scholastic ideas. In addition, a new rationalist thought begins to develop and to create a new context for science development. Progressively, caused mainly by cartesianism, but not exclusively by these ideas, this thinking had expanded and become popular, going beyond scientific circles and contributing to changes in the established vision of nature.[6]

The machinism expansion, which characterizes modern industry, has replaced the previous form of labour and originated a new process of production no longer dependent on artisanal skills and qualifications but on machine characteristics such as velocity, regularity and continuity.[7] If we compare with manufacturing, mechanized production has increased the contribution to product in terms of aggregated value. Labour productivity increased by a machine can be measured by human labour force replaced by the machine itself. Another consequence of machinist expansion is price lowering and diminished human labour that is replaced by use of machines. Thus, capitalists were paid by labour-force utilization and not by labour itself and thus the limits for the use of machines by capitalists are calculated by the difference between machine value and labour-force value that is replaced. It is necessary to emphasize that machine expansion in capitalist society has the objective of increasing labour productivity, becoming the driving power of the creation of relative surplus-value.[8]

In order to increase the surplus-value of the expansion of machines in a productive system it is necessary to promote the general lowering of merchandise prices, which means a decrease of the labour-force value by a decrease of the number of workers associated with an amount of capital. In other words, variable capital must be transformed into constant capital. This necessity is the essence of capitalism dynamics and the continuous development of machines.[9]

---

[6] About the relations between science and the manufacturing period, see (Labastida 1978; Alquié 1987).

[7] Marx analyzes in the following way the introduction of machinism in this new phase of capitalism: "its utilization (of machines) as any other development of productive force, has the objective to decrease merchandise prices, to decrease the journey of labour which workers need to themselves in order to increase the other part which is given to capitalist. Machinery is a mean to create surplus-value". (Capital, Vol. I, Chap. XIII, p. 424 (Marx K (1968)).

[8] The concepts of absolute and relative surplus-value are defined by Marx as follows: "The production of absolute surplus-value is done with the increasing of labour journey beyond the point where worker produces only the equivalent to the value of his labour force with the appropriation by capital of this amount of exceeding labour. This process is the essence of capitalist system and the starting point of the relative-surplus value production. The last supposes that labour journey has been already divided in two parts: necessary labour and exceeding labour. To increase exceeding labour necessary labour is shortened with methods that permit to produce in less time the equivalent to salary. The absolute surplus-value production is then associated to labour journey duration; the relative surplus-value production supposes a revolution in technical processes of labour and social aspects". (Capital, Vol. II, Chap. XIV, p. 585 (Marx 1968)).

[9] About the definitions of constant and variable capital see "organic capital composition" in p. 69, Ref. (Bottomore 1988).

This socio-economic context explains the conditions for development of the first general theory of machines proposed by Lazare Carnot at the beginning of the nineteenth century, as well as how an industrial mechanics could appear by the work of polytechnician engineers. A new theory of machines was required as a necessity of the industrial processes in France and in other European countries. The questions related to efficiency, optimization of human and material resources, costs, etc. are of great relevance to economic production and also stimulated the development of that theory as a previous condition for the solution of these complex problems.

Lazare Carnot's theory was built within the conceptual framework of Rational Mechanics, at that moment completely developed and including Lagrange analytical mechanics which appear in 1788. What is important to remark is the geometrical character of Lazare Carnot mechanics based on d'Alembert's principle. His mechanics differs substantially from Lagrangian approach and studies particle dynamics where shocks and sudden changes of speeds and trajectories are common, unlike the Lagrangian approach. Like Lagrange's theory, Lazare Carnot's theory uses differential and integral calculus, which was a recent development by several mathematicians.

Machinism Effect on Labour: Fatigue

If we consider fatigue from an individual point of view, we observe that a worker isolated from the overall production process represents many problems, which also characterizes industrial machinism. To maintain workers' physical and mental equilibrium, it is necessary to know each particular job as well as some biological and psychological information, mainly body limits and the causes that can change that equilibrium.[10]

With the Industrial Revolution, and the progressive applications of scientific knowledge to economic production, it was not taken into account that the new technical labour conditions were very far from promoting a state of satisfaction to workers. At the beginning of the Industrial Revolution, the well-known first Industrial Revolution, old crafts continued with the unity of labour process. This situation has now changed completely. New forms of labour division have appeared and with it decomposition into multiple operations such as automation and serial production, all of these aspects submitting workers to different conditions, physical and psychological. In addition a new kind of influence such as programmed rest, interest in welfare of labour, technical education, specializations, skills, qualifications, etc., indicate the necessity of a new approach to labour.

At the beginning of the twentieth century appears what we can call an industrial science of labour. Numerous studies, books, reports, etc. were published in correlated

[10] Although an old book it is a very interesting study of problems created by machinism for human beings. See Ref. (Friedmann 1946).

fields such as physiology, psychology, a sociology of labour, ergonomics and a science devoted to labour organization. In this context the scientific administration proposed by Frederik Winslow Taylor (1856–1915) arose.[11] This new discipline intends to be the symbol of industrial modernity. What is important to remark is the emergence of a labour science from the end of the nineteenth century to the beginning of the twentieth century. In this same period social and economic transformations were responsible for the development of modern industrial companies (Vatin 1999).

To understand the origins of fatigue studies, as we have seen to be the most characteristic effect of machinism, we should refer to the pioneering work of Charles Augustin Coulomb (1736–1806) about fatigue. He uses mathematical tools of Rational Mechanics to develop several new concepts. This was done in his *Memoire sur la force des homes*, published in 1778. Concepts like network, total work and mechanical efficiency were introduced and this can be seen as the birth of applied mechanics. These new concepts were progressively incorporated into this new discipline and later on into economical thought. The participation of other engineers and mathematicians like Navier, Coriolis and Poncelet will be studied in detail in Chap. 7 (Vatin 1993).

Historically, the approach used by Coulomb to study fatigue problems with a mechanical model has an important goal, which is to consider the old question of how to measure human capacity to work during a journey. His basic concern was to measure mean values of work in a journey and not maximum or minimum values. In other words, Colomb was looking for an actual measurement of work in a regular journey. Frederik Taylor will make the same considerations one century later.

In the Coulomb model it is necessary to distinguish two factors in the work of men or animals: the effect that can be obtained using mechanical energy to drive the machine. The second factor is the fatigue associated to this effect. Coulomb proposes a mechanical model to maximize the ratio between the effect and the fatigue, i.e., the efficiency. This process indicates an economic concern and is in the origin of applied and industrial mechanics as scientific disciplines.

The theoretical model proposed by him analyzes a man displacing a weight in vertical and horizontal motion. For the sake of simplicity and as an illustration we consider only the first motion. The effect that one can obtain is the product of the weight by the height to be reached by weight. To obtain total work it is necessary only to add a man's weight. The approach applied is explained using the following nomenclature:

P = weight to be elevated

Q = man's weight
h = elevation height
$(P + Q)h$ = quantity of action
Ph = net effect

[11] The most important works of Taylor are: A Piece Rate System (1895) and Shop Management (1902).

Coulomb performs a detailed analysis of the elevation problem which is avoided here. He considers the quantity of action, the same as work, in one day of labour but fatigue is taken into account. The model representing fatigue is a function $(a-bP)$, where the parameter a is the maximum quantity of action in a day and b is a quantity representing the negative effect of the weight. Obviously, he uses a linear model for the sake of simplicity. He equates: $(a-bP) = (P + Q)h$ or $Ph = a-bP-Qh$; this is the function obtained by isolating the net effect as a function of the parameters a, b and the load P. His model is a function to be studied by differential calculus rules. In fact Coulomb should minimize the losses and find the values of the parameters which imply this minimization. He finalizes the analysis with a discussion about the numerical values that can be attributed to parameters a and b and the physical interpretations derived from the model. What we should explain is that the origins of the concepts of network, total work and efficiency can be easily identified in Coulomb's analysis by mechanical analogy. The word work is not yet used here.

Concrete and Abstract Labour

Merchandise, which is a product to be interchanged in a capitalist market, appears as a unity under two different aspects: its utility to customers, giving a sense of interchange and the capacity to obtain quantities of other merchandises in the interchange process. The first aspect is called use–value and the second exchange–value.[12]

Although, the use–value could be a necessary condition such that the product can be exchanged performing the exchange–value, such that the first one has no quantitative relation with their corresponding exchange–value, leaving that correspondence to depend on merchandise production. As a consequence, the main objective of political economy is the knowledge of governing laws of production and the dynamics of exchange–value, which is the same as knowing the governing laws of value, the properties of merchandise production and how the exchange–value appears. This fact explains why the great majority of labour studies deals with exchange–value. In a Marxist terminology it means abstract labour.[13]

---

[12] At the beginning of "Capital" Marx defines merchandise as: "Merchandise is initially an extern object, one thing that, by means of its properties, satisfies human needs of any nature and origin from stomach or from fantasy". (Capital, Vol. I, Chap. I, p. 41 (Marx 1968)).

[13] Marx defines abstract labour as follows: "Forgetting for a while use-value, and abstracting the forms and material elements which characterizes it as use-value, we can no longer see a table, a house, a string or another useful thing. All the material qualities have disappeared. In addition the product is no more a cabinet-maker labour, or a mason labour, or else a spinner labour, or any kind of productive labour. When the utility character of products has disappeared at the same time, the utility character of incorporated labour also disappears and the different forms of concrete labour cannot be distinguished and become a unique kind of labour, human labour abstract". (Capital, Vol. I, Chap. I, p. 45 (Marx 1968)).

From a general viewpoint, even using a Marxist analysis, several important questions should be explained regarding merchandises and associating them to concrete labour, the Latter being considered as a material substrate of exchange–value.

Roman Rosdolsky (1898–1967) highlights this fact in his famous book. It was published as: *Genesis and Structure of Karl Marx's Capital* (Friedmann 1946). We can find a whole chapter studying use–value with the title: *Karl Marx and use–value problems in political economy.* Rosdolsky has discussed this problem in order to eliminate some mistakes from important Marxists' authors such as Rudolf Hilferding (1877–1941) and Paul Sweezy (1910–2004). He states: *Before to explain Grundriss's content, we shall consider a methodological problem until now underestimated by Marxist literature whose solution depends strongly on Grundrisse's knowledge. This problem is the role of use–value in Marx's economy.*[14]

Rosdolsky quotes Marx (1818–1883) from his masterpiece as follows: *As we know labour process involves man and nature, then its simple elements are common to any social forms of development. But, each historical form of this process mainly the specificities develop their material basis and social forms.* Rosdolsky continues: *Nature laws cannot be neglected. In different historical circumstances, what can be modified is the form under which these laws appear.*

In this investigation we intend to preserve the fundamental unity of man–nature, analyzing the specific forms where nature laws, in particular physical laws, were used to characterize use–value in the Industrial Revolution framework spreading in Europe. To this accomplishment it is necessary to equilibrate our analysis by means of questions from different fields of physical and social sciences. The interdisciplinary character of our investigation does not limit the analysis of the work concept because it is, of course, responsible for the transformations on use–value. Indeed, these transformations are a unique process where the value to be attributed to merchandises are given by its exchange–value. When the living forces principle is applied to machines, this fact itself amplifies our field of analysis because a machine is a social artifact.

An interesting example of analysis which articulates elements from natural and social sciences is presented by Enrique Leff in his book *Environment Epistemology* (Leff 2002). Leff's book makes a very interesting analysis using biological concepts and their relations with ecological questions. Leff states: *In history science nature appears as labour objects and potentialities from nature which integrates the whole process of capitalist production and, in general all productive processes encompassing a social transformation, is an effect of the reproduction/social transformation process.* He continues: *The absorption of nature by capitalist process of production where nature is represented by labour objects, resources and natural phenomena or even ecological productivity, all of them are incorporated*

[14] The natural form of merchandise, what Marx denominates as use-value, does not interfere directly in the process of value formation, "it is only the material substrate, the mobile of exchange-value". Here, in addition it means the philosophic dimension of Marx and the production process should be considered in its historical meaning as productive labour process of use-value" (Rosdolsky 2001).

technologically to productive process. Leff finalizes with the following remarks: *A general science of history does not exist, consequently no one general concept or notion can be used as scientific tool to articulate nature and society. Labour process that means a transformation of labour objects in use–value is the general condition of all mode of production. As result the concepts of labour or use–value cannot explain specific determinations within labour process belonging to a determinate mode of production as well as their consequences in nature transformations.*

The quotation above agrees with our analysis mainly when we emphasize the importance of use–value in the same sense used by Rosdolsky. The difficulties of the majority of analysis are neglecting the central role of the capitalist production, taking into account the fact that the capitalist mode of production alone cannot explain specific forms, and how labour processes can be developed as well as their consequences. It is worth remarking that our investigation deals with specific forms such as mechanical principles and how they were used to analyze machines regarding productive processes. Our contribution should be accepted as an attempt to equilibrate the other side of merchandise, the use–value side. In other words, this analysis avoids the duality use–value/exchange–value. In fact, this duality does not exist in Marx's theory. He separates both concepts from a didactic point of view without dualism.

In order to complete our considerations about merchandise in general, let us see how abstract labour is used as a value norm. Russian economist Isaak Rubin in his *Marxist Theory of Value* (Rubin 1987) states: *To understand thoroughly Marx's theory of abstract labour, it is worth to comment that Marx associates the concept of abstract labour with the concept of value. Abstract labour "creates" value and is the "content" or value "substance". Marx's task was not (as frequently we can see) to reduce the value, analytically to abstract labour but to derive the value in a dialectical way value from abstract labour as starting point. This cannot be done if abstract labour is considered as labour in physiological sense.* Rubin continues: *Physiological labour is a necessary condition to abstract labour existence, so that we cannot speak about abstract labour without a loss of energy from physiological activity. This loss of energy stays as a necessary condition but not as the objective of our analysis.*

Summarizing the above considerations it is important to conclude that both types of labour, abstract and concrete, can be separate only in terms of investigation, but any kind of dualism is meaningless and they belong to the same reality. In other words, concrete labour as a necessary condition and abstract labour as a historical concept both pertain to a capitalist stage of development, the stage of merchandise production. In this stage the form to attribute value to merchandises is characterized and specific to these social relations. In case of a profound capitalist transformation and its substitution by a new and different mode of production, socialist or not, the situation can change. Finally, using the denomination of abstract to designate labour is due to the fact that this activity abstracts material and physical aspects and it became concrete by means of a well–defined social relation historically determined.

Labour and Value

In Chap. 7, we will show the historical process where the concept of work is incorporated into the economy. It is also shown that in the context of machines all attempts to use the concept of work as a value measurement, presents similarities with Marx's theory of labour–value. What is common in Marx's theory with the approaches of Navier and Coriolis is to find a rule or a kind of measurement to establish value of the product as well as to measure machine consumption such that it can be quantified to compare productivity and efficiency of any machine. In terms of economy the main question under consideration in Marx's theory is the same weakness of exchanges in economic theory presented by classical economy.[15] Exchanges are based on equivalence principle. How then can we understand that in several exchanges there appears a surplus value which explains economic growth as well as capital accumulation?

In the theories of Navier and Coriolis, as well as in Marx's concept of abstract labour, in a second step, after obtaining a norm of value, these theories also search to provide how to quantify costs and production expenses. To Marx a productive expense is composed of two fractions: merchandises already quantified in terms of work encompassing production means, such as machines, raw materials, etc., and what Marx calls dead work whose value is only incorporated into a new product value.

Thus, it is possible to use abstract labour to measure product value as production expenses which are labour–force value. These two quantities imply a surplus–value which can be quantified in terms of labour–value: the surplus–value and a ratio expressing economic efficiency of the process. In other words, capitalist gain. The ratio is defined as the quotient between surplus–value and expenses.

It is easy to see the analogy between Marx's construction and that of polytechnical engineers, mainly Navier and Coriolis. Labour is used by both theories to measure the product and its cost which permits us to express the whole process in terms of value, and then a relation numerator/denominator with the same dimensions. In Marx's theory a surplus–value tax and in Navier's and Coriolis's

[15] About the weakness of classical economy to analyze the value problem, Marx said: "Political economy has analyzed, in fact, in an incomplete way, the value and its magnitude and discovered the hidden content. However never ask it why this content was hidden, why labour is represented by value of labour product and the duration of labour time by the magnitude of this value". (Capital, Vol. I, Chap. I, p. 89 and 90 (Marx 1968)). In note 33 of the translation one reads: "Classical political economy fails to analyze merchandise, especially merchandise value, - value form and how it became exchange-value. The most famous representatives as Adam Smith and David Ricardo, treat with absolute indifference the form of value or have considered this form not correlated to merchandise nature…".

approaches, efficiency as a ratio between net work/total work. In spite of similarities a fundamental difference appears between these two theories.[16]

The mechanical model used by polytechnical engineers express a value and a loss. The product value is expressed in terms of net work that is lesser than expense which is total work and mechanical efficiency. In Marx's theory an inverse situation appears that is a surplus–value and the possibility to have product value greater than expense. Thus, a labour–force can produce an amount of labour superior to its cost. This is a peculiarity of a "special merchandise" labour–force, the only entity that can create value. This aspect is the fundamental difference between Marx's theory and that of the polytechnical engineers. For a given machine, thermodynamic laws imply the impossibility to have mechanical efficiency greater than or even equal to one. This fact also shows that it is nonsense to attribute to Marx an energeticist conception which considers human labour equal to machine work. In this case surplus–value would be a negative quantity and a loss instead of surplus–value. Friedrich Engels (1820–1895) stated that there is a big mistake in Sergei Podolinski's (1850–1891) paper in which he thought to create a social theory based on pseudo–Marx energeticist theory of value.

It is important to remark that after discovery of the energy conservation law in 1847, a theoretical conception based in this law was developed. The search for a unified knowledge had influenced social sciences. The energy concept itself provides arguments to give to energy an economic interpretation. Energy is like "currency" that can be exchanged in any field of physics. As proposed by Navier: a mechanical currency.

On the other hand the concept of energy is related to human activities in several ways in economic production using mechanical forces and as biological specie needing some amount of energy for surviving. This general aspect of energy and its applicability can create a false feeling that energy could provide a common measurement to any good, merchandise or not. This is not Marx's conception of value created by labour in merchandise production.

[16] About the analogies between the theories of Navier and Coriolis on one hand and Marx's approach, Vatin (Vatin 1993) wrote: "The analogy with mechanicians construction is very strong. Work is a common measurement to product and expense, which permits to use a value relation where numerator and denominator are quantities of same dimensions (calculated by the same measurement): tax of surplus-value to Marx, efficiency (net work over total work) to mechanicians. However, at the same time a fundamental difference between two constructions appears clearly. Mechanicians express a value and a loss; product value (net work) is necessarily lesser than expense and efficiency is then assumed less than one. Marx on contrary express in value a surplus-value supposing that product value can be superior to that of expense; labour force can produce more labour than its cost".

Nature Sciences Versus Human Sciences

Nature studies are deeply related to human relations because man cannot develop an understanding of nature in a pure form outside society but integrated with the human world by means of labour. This implies that our knowledge about nature is an anthropologic knowledge, (Vasquez 1968; Marx and Engels 1975). [17], [18]

To Marx nature sciences are human sciences and the sense of his analysis is that man is the immediate object of nature sciences and nature is the immediate object of man. Thus, to Marx there is an inseparable unity between these two fields of knowledge. Marx foresees for the future a fusion between the two fields because of the common anthropological basis.

Marx also looks at the division between man and nature that implies a division between nature science and human sciences as being due to an alienated relationship. This concept is related to the separation between man and the products of his labour. These ideas are presented and developed in Marx's texts about the relation man–nature and we will use this framework in our analysis.

An example of an inextricable relation between nature and human sciences is the evolution of the scientific–natural model of objectivity. Nature sciences have been acquiring a progressive freedom from ideological and value judgments from a very long historical process. During feudalism, weakness of the dominant classes from the political and military points of view increased the fight against enemies in the ideological field in order to maintain the system of domination. The ideological system of established values was composed by dogmas and in complete agreement with static and immutable order of the universe. Hence, it is easy to understand why any controversy or divergences from that order even through nature sciences ideas were punished with rigor and violence as heretic manifestation. Thus, the examples of Giordano Bruno (1548–1600) and Galileo illustrate this situation. The political and ideological struggle occurred in the nature sciences field.

With the birth and development of the capitalist mode of production, discussions on nature sciences were losing their ideological characteristics. This happened because capital in general, spurred on by the Industrial Revolution, needed scientific knowledge, whereas conflicts and ideological disputes presented a displacement of

---

[17] This anthropologic view that Marx has of nature can be found and discussed in details in Adolfo Sanchez Vasquez book (Rubin 1987) where one can read on page 143: "The problem being-object is formulated in *Manuscripts* where Marx studied the relations between man and nature. In *Feurbach´s thesis* this problem is studied as a problem being-object. In each plane Marx reaches to the same conclusion: the second term of the relation—nature, in *Manuscripts*; object in *Thesis*—does not exist to man isolated from practical activities and, thus has to him an anthropologic character. This anthropologization of nature and of object—character incorporated by practices—determines an anthropologic characterization of knowledge, that is, of cognoscente's being-object nature".

[18] In the introduction of Alfred Schmidt's book (Schmidt 1994) he comments on Marx's vision about nature: "This work is a contribution to a philosophical interpretation of Marx. He shows an interest by a secondary concept in other places of his work that is the concept of nature. In Marx writings rarely appear the nature *by itself*. This is a particular point of view that Marx adopts".

social and economical questions from which are derived the products generated by labour. As soon as the capitalist mode of production was settled in the majority of cities in Europe, at the end of the eighteenth century and beginning of the nineteenth century, nature sciences were emancipated and dissociated from a religious ideological basis. It is not coincidence that this period is characterized by preeminence of illuminist philosophers who addressed their critiques against religious feudalism, the authority principle, and scholastic dogmatism, which was a fundamental step toward transformation of the scientific–natural model of objectivity.

Later on, this model matured and a new epistemological ideal based on science free from ideology and values judgments appeared as a kind of neutral science. This ideal, taken to its extreme, created the positivist scientific model (Löwy 1987, 1975).[19]

Nowadays ideological questions continue to arise in nature sciences in a transformed way. The selection of research themes, technical applications and discoveries depend on economic and social group's interests that have control of investments and appropriation of gains.

If we analyze the model of scientificity where fields of knowledge are separated, a comparison between nature sciences and human sciences can show that there is no absolute difference between them. In addition there exist no well-defined areas, fields of transition and interfaces between both types of science. In these cases we can find ecology, some parts of biology, compared psychology, etc. In these fields both nature and human knowledge are required.

With respect to human and social sciences, unlike the thought of Auguste Comte and the positivist School, it became impossible to separate scientific knowledge from value judgments. Michael Löwy uses a physical metaphor to characterize this problem. He said that when a nature science approaches the border which separates it from human sciences, there is a kind of "ideological heating" and it becomes electrically charged.[20]

In spite of this epistemological impossibility, positivist ideas persist within their analysis and in their model of objectivity. They deny the difference between two fields and postulate a similarity between nature and social laws.[21]

In general some methodological differences between nature and human sciences do exist and can be presented as follows:

(a) Historical character of social and cultural phenomena and the possibility they could be transformed by men's actions in a different way than happens with

---

[19] See Michael Löwy's book "The adventures of Karl Marx against Munchausen Baron" (Löwy 1987) and "Dialectical Method and Political Theory" (Löwy 1975).

[20] To read a critical vision about positivism see Herbert Marcuse in *Reason and Revolution* Marcuse 1978). See also (Benoit 2002; Kolakowski 1988).

[21] In page 12 of reference (Löwy 1975) one can read the following quotation of Auguste Comte: "I understand by social physics the science that has by object social phenomena, considered within the same spirit as the astronomic, physical, chemical and physiological phenomena, that means, they are subjected to invariable natural laws, whose discovering is the especial objective of these researches".

nature laws where men's action's have the objective of knowing them to be used.
(b) An identity between being and object.
(c) Dominant classes had interpreted past and present, social and political conflicts according to their life and social experiences and interests. Consequently historical sciences are inseparable from value judgments.
(d) Knowledge and/or its recognition can have profound consequences on social classes' behavior. It can modify political forces correlation. Then, to reveal or to hide the truth is a political weapon in the class struggle context. Antonio Gramsci (1891–1937) stated that *the truth is ever revolutionary.*

The reasons enumerated above show that methodologies used by both fields of knowledge are quite different. As a result, models of objectivity are also different. In Chap. 1 we will analyze the question of ideology and value judgements and argue that it makes the difference between the two scientific fields.

References

Alquié F (1987) Galileu, Descartes e o Mecanicismo. Gradiva, Lisbon
Anderson P (1991) Passagens da Antiguidade ao Feudalismo. Editora Brasiliense, S. Paul
Barret F (1975) Historia del Trabajo. Editorial Universitária de Buenos Aires, Argentine
Benoit LO (2002) Auguste Comte: Fundador da Física Social. Editora Moderna LTDA, S. Paul
Bottomore T (1988) Dicionário do Pensamento Marxista. Jorge Zahar Editor, Rio de Janeiro
Bowler PJ (1989) Evolution: the history of an idea. University of California Press, Berkeley
Friedmann G (1946) Problèmes Humains du Machinisme Industriel. Gallimard, Paris
Jaccard P (1974) História Social do Trabalho. Vol I. Livros Horizonte, Coleção Movimento, Lisbon
Kolakowski L (1988) La Filosofia Positivista. Ediciones Catedra S. A., Madrid
Labastida J (1978) Produccion, Ciência y Sociedade: de Descartes a Marx. Siglo Veintiuno Editores, Mexico
Leff E (2002) Epistemologia Ambiental. Cortez Editora, S. Paul
Löwy M (1975) Método Dialético e Teoria Política. Editora Paz e Terra, Rio de Janeiro
Löwy M (1987) As Aventuras de Karl Marx contra o Barão de Munchausen. Editora Busca Vida, S. Paul
Marcuse H (1978) Razão e Revolução. Editora Paz e Terra, Rio de Janeiro
Marx K (1968) O Capital. Editora Civilização Brasileira, Rio de Janeiro
Marx K, Engels F (1975) Cartas sobre as Ciências de La Naturaleza y las Matemáticas. Editorial Anagrama, Barcelona, Spain
Rosdolsky R (2001) Gênese, e Estrutura de O Capital de Karl Marx. Editoras UERJ e Contraponto, Rio de Janeiro
Rubin II (1987) A Teoria Marxista do Valor. Editora Polis, S. Paul
Schmidt A (1994) Le Concept de Nature chez Marx. Presses Universitaires de France, Paris
Vasquez AS (1968) Filosofia da Praxis. Editora Paz e Terra, Rio de Janeiro
Vatin F (1993) Le Travail: Économie et Physique 1780–1830. Presses Universitaires de France, Paris
Vatin F (1998) Économie Politique et Économie Naturelle chez Antoine–Augustin Cournot. Presses Universitaires de France, Paris
Vatin F (1999) Le Travail, Sciences et Societé. Éditions de l'Université de Bruxelles, Brussels

Chapter 1
Theoretical Framework

The original purpose of Historical Materialism was to present a theoretical basis to interpret the world in order to change it. This was not only a slogan. It had a well defined meaning. It also means that Marxism has looked for a special type of knowledge able to highlight principles of historical movement and, at least, implicitly, the points where political action could to promote a transformation with more efficacity.

(Ellen Meiksins Wood—Democracy against Capital—Boitempo Editorial, S. Paul, 2003).

1.1 Marxism

1.1.1 A Brief History (Hobsbaum 1980)

Karl Marx (1818–1883) and Friedrich Engels (1820–1895) were Marxism's founders, both born a decade after the Napoleonic wars. Marx was the son of a lawyer from Trier, and Engels the son of a rich industrial man from Barmen. This means that both had a Rennin origin. They lived in prosperous families from a well developed German region.

From his youth Marx was interested in insurgencies by European workers in the wake of the Industrial Revolution, and between the ages of twenty and thirty he came to a kind of settling of scores with the philosophical legacy of Hegel (1770–1831) and Feuerbach (1804–1872). Engels, living in England, discovered the reality of English working class life and fought against all political doctrines that legitimated the *status quo* (Fig. 1.1).

As a consequence of their political engagement in the English working class movement, they collaborated on a programmatic document, entitled *Communist Manifesto*, published in London in 1848. This happened on the eve of a political conflagration in Europe. Consequently, the counter-revolutionary movement obliged Marx and Engels to exile themselves in England. Marx made a political

A. R. E. Oliveira, *A History of the Work Concept*, History of Mechanism and Machine Science 24, DOI: 10.1007/978-94-007-7705-7_1,

Fig. 1.1 **a** Karl Marx (1818–1883). **b** Friedrich Engels (1820–1895)

balance of the French Revolution and Engels was dedicated to understanding the reasons why the German Revolution failed.

Marx, living in extremely difficult conditions in London, decided to analyze the capitalist mode of production globally. After 15 years of hard work he published the first volume of *The Capital* in 1867. Simultaneously he participated in the foundation of the first International, the first workers organization of international scale. In 1871, he celebrated the eruption of *Paris Commune*, an insurgency of craftsmen and workers which managed to seize power for some weeks in Paris. *Paris Commune* inspired Marx to future theoretical works about a new kind of state, a worker-state.

In the last years of his life and even some years after his death, Engels produced the first systematic publications on historical materialism, trying to popularize the theory created by himself and his partner. At the age of seventy Engels promoted the foundation of the second International and assisted Marxist theory in becoming the official doctrine of the majority of workers parties in Europe. The Socialist parties arose in this period and the new International significantly changed his political configuration, becoming a kind of party federation.

Some characteristics of Marx and Engels' social work should be emphasized in order to understand its real meaning and future developments. The first

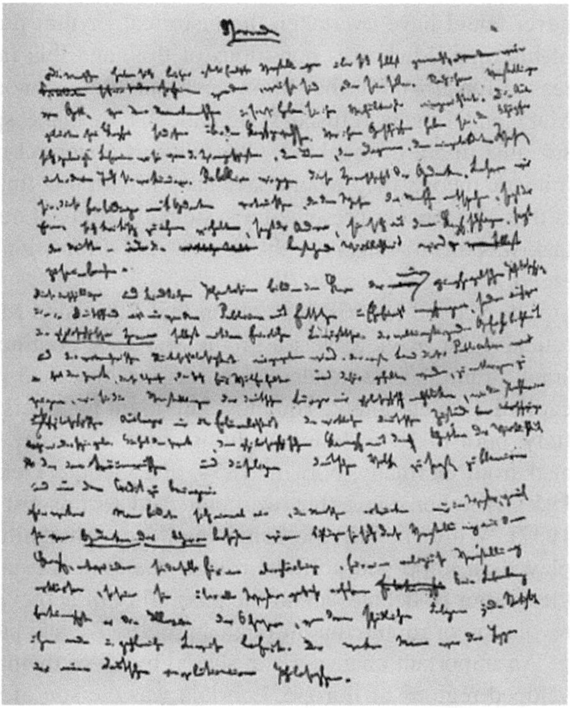

Fig. 1.2 A page of German Ideology Manuscript

characteristic of this fruitful cooperation is that they were pioneers and isolated members of their generation. No one in any country or of any nationality participated in or understood their concepts. The monumental work that both built was result of a long partnership across four decades, approximately (Fig. 1.2).

Another important characteristic of this work is that they both lived in extreme deprivation, especially Marx. They suffered exile, persecution, and fatigue, which tested their solidarity and friendship. In some cases, they did not participate directly in the workers' struggle of the time. This analysis highlights the existence of a complex articulation between the theory that they developed and its practical political engagement. In conclusion we can say that limits of their theoretical work were the limits of the worker's movement of the given time. Another conclusion is that the influence of Marx's theoretical work was minor in his lifetime, with the majority of his works being published five decades later, approximately.

After his death, Marx's legacy was a well-developed and consolidated economic theory analyzing the capitalist mode of production that he had begun in his works: *Economic and Philosophical Manuscripts, The Poverty of Philosophy and Contribution to Political Economy Critics.* This happened a decade before the publication of the first volume of *The Capital.* On the other hand Marx neither developed a political theory for the state nor tactical and strategically political programs towards a socialist society. With respect to Marx's theories we can say that they never could have overtaken the historical rhythm produced by social movements of his time. Marx was conscious of this, and this fact is a good epistemological reason for affirming that Marx never adopted economic determinism as doctrine. Marx and Engels refused systematically to make speculations or to predict the socialist future of mankind. Otherwise, if historical materialism had been a deterministic theory, they would have used it with that finality. Another important point is that Marx never presented a systematic study about historical materialism. This task was left to Engels in the works: *Anti-Duhring* and *Ludwig Feurbach and the end of Classical German Philosophy*.

The Second Marxist generation that succeeded Marx and Engels, from a theoretical point of view, was a group who gave continuity to Marx's project. It was small in number and ended up bowing to historical materialism in a late phase of capitalist development. The most important members were Antonio Labriola, from Italy, born in 1843, Franz Mehring, from Germany, born in 1846, Karl Kautsky, also from Germany, born in 1854, and George Plekhanov, from Russia, born in 1856. Another name that we could add to this list is Eduard Bernstein (1850–1932). While not of the same intellectual magnitude as the others, Bernstein played an important political role as the first to examine the theory's history and the spurrer of debates about the possibility of achieving a socialist society without revolution or rupture but through a progressive and pacific way.

An important characteristic shared by all of them is that they came from undeveloped regions of Europe. Labriola was the son of a landowner from Campania. Mehring came from Prussia, Kautsky from Bohemia and Plekhanov from Russia. Although this group had a general political importance in their respective countries, only Plekhanov had participated in a revolutionary movement as a member of the Central Committee of the Russian Social Democratic Labour Party, the matrix of the Bolshevik Party.

All of them maintained personal contact with Engels and he influenced their works. Each by his own methods attempted to give a systematized structure to historical materialism so as to face it as a general theory of both society and nature. They tried with their works to create a coherent world vision in order to give workers movements a powerful tool to analyze and transform their political situations. They also fought to replace all academic disciplines that opposed Marxism with historical materialism. Franz Mehring wrote: *On Historical Materialism;* Antonio Labriola: *Essays on Materialist Conception of History;* Plekhanov: *The Development of Marxist Conception of History* and Kautsky: *Materialist Conception of History.*

In general these works served the purpose of complimenting Marx's theoretical effort. It is important to remark that Engels published volumes two and three of *The Capital* after Marx's death while Kautsky published *The Theory of Surplus-Value* (based on the work Marx had done for volume four) and Mehring worked on *Marx and Engels Correspondence*. It was also Mehring who published the most important biography of Marx shortly before his own death.

A more deep vision of the intellectual production of this generation of Marx's disciples is necessary to understand their contribution in its real context. The situation at the end of the 19th century was characterized by accentuated economic growth in the majority of industrialized countries, by strengthening and stabilization of economic monopolies associated with a great imperialist expansion. As result, the first generation of Marxists after Marx lived in a relatively unperturbated political climate with economic growth. It is easy then to understand why a segment of these social democrats tried to interpret their social and political reality through a gradualist vision owing to the economy of this period. Apparently capitalism would be developed continuously with the possibility of distributing products of this development with the dispossessed. In a more successful perspective a socialist society would be the crowning of this process. Obviously, this analysis is not complete and other influences have to be considered to explain this historical period. In this context the gradualist ideas of Eduard Bernstein could be presented as political path to socialism. In other words Bernstein had denied class struggle and any revolutionary rupture.

The third generation of Marxists, the second after Marx and Engels, would find their entire situation quite different from the previous one. European capitalism became accelerated following the First World War. In addition this new generation was larger and subject to different experiences as well as a new political context. The main theoreticians were Lênin, son of a public employee, born in Simbirski on the Volga River in 1870, Trotsky, son of a farmer from Ucrania, born in 1879. Bukharin (1888–1938) and Preobrazhenski were Russian. The last three Marxists participated in the Russian Revolution alongside Lênin and are considered famous theoreticians of the Bolshevik party.

This new context would guide the theoretical works of this last generation of Marxists. Thus, one of them deals with changes in the capitalist mode of production, in an attempt to update and continue *The Capital* of Marx. *The Agrarian Question* written by Karl Kautsky, published in 1899, tried to explain agrarian transformations that had occurred in Europe and United States as result of capitalism development. Other important publications included *The Development of Capitalism in Russia* by Lênin and, finally, the book most closely aligned with Marx's *Capital*, *The Financial Capital* written by Rudolf Hilferding, published in 1910. Hilferding was a great theoretician of the Marxist economy and came from Austria. He taught economics in political schools maintained by the Social Democratic party. He was succeeded by Rosa Luxembourg (1871–1919).

It is important to remark that the inversion in the political situation of the period of the second generation of Marxists meant to them the acceleration of capitalist crisis and in this context the birth of the October Revolution. This situation would impact theoretical production significantly.

With the above mentioned inversion in the socio-economic situation the political analysis also changed and it would be natural for works to arise predicting and anticipating the Revolution in Russia. To Marx it would be a heresy because, according to him, mature conditions in the revolutionary process would appear only after some degree of capitalist development. This was not the situation in Russia. The great theoretician of this situation was Leon Trotsky who, several decades before the Russian Revolution, had developed the theory of *Permanent Revolution* which predicts and justifies a socialist revolution in undeveloped countries. From a political point of view this theory was adopted by Lênin when he arrived in Russia in 1917 and wrote a famous pamphlet entitled *April's Theses*, astonishing his party's comrades. They accused Lênin of Trotskyism.

Lênin justified this change in his political program, stating that an imperialist chain could be broken at its weakest link, which would be Russia. This rupture with the Marxist orthodoxy would divide the revolutionary movement into two big blocs: Mensheviks and Bolsheviks. The last bloc standing would be the revolutionary flag of socialist revolution in Russia and be responsible for seizing power over the nation (Anderson 1976).

1.1.2 Historical Materialism

Materialism postulates that the real world is constituted and maintains a subordinate dependence on matter. In the philosophical plane the division between materialists and idealists depends on how a philosophical school explains the dependent relation between being and thought. Materialists are those who defend the subordination of thought to being while idealists defend the inverse relation.[1, 2] Obviously, these two categories are related to the philosophies of materialism and idealism. From the point of view of the philosophy of sciences, modern physics redefined the concept of matter.[3] Einstein showed in his famous papers published in 1905 (Bodanis 2000; Stachel 2001) that matter could be converted into energy and vice versa. When we speak of matter, we mean a condensed form of energy. In the next Chapter we will discuss fundamental aspects of matter regarding classical mechanics. Thus, the concept of

[1] For the purposes of our work there is no need to study matter in depth. In addition we restrict our investigation to the field of classical mechanics and thus matter is the same as mass, being one of the fundamental concepts of mechanics. We will come back to the concept of mass in the Chap. 2.

[2] Matter, from the philosophical point of view, is a complex concept. Lenin (1971) remarked that everyone that discusses questions related to matter always falls into the old dilemma between being and thought, stating: "The reader sees that all arguments from the empirocriticism´s founders turn exclusively around the old gnosiologic problem of relations between thought and existence, between the sensations and the physical …". Later on he defines matter as follows: "The agreement with philosophical trend denied by idealists and agnostics express itself in the following definitions: matter is what is, acting in our feeling organs, and thus creating sensations; matter is an objective reality which is gave to us by sensations, etc. …". He continues: "Matter is primary fact and spirit the second fact to a philosophical tendency, while to other one, the inverse".

[3] From the physical point of view, what is now accepted about matter is that it can show itself as follows: visible matter, or common matter, composed mainly by protons and neutrons; dark baryonic matter, very thin common matter to be seen; dark non-baryonic matter, a kind of exotic particles like *axions* or neutrinos without mass; dark energy which is the energy of empty space.

mass will be studied in this new context and within the possibility of the discoveries of Higgs boson.

From a terminological point of view philosophical materialism can be divided as follows:

(a) Ontological materialism, which states the dependence of social being on biological being and the emergence of the first from the second;
(b) Epistemological materialism, which states the existence of the physical world and its independence from objects of thought.
(c) Practical materialism, which believes in the transformative action of man in his social organization.

From a historical point of view to think about ancient materialism is to think like Epicurus (341–270 BC).[4] Another important figure in the history of materialism is Democritus (460–370 BC) (Marx 1972) whose work was practically lost while Epicurusan manuscripts are numerous. On the other hand Lucretius (95–55 BC),[5] who left many texts and manuscripts, is also considered a doxograph, a kind of compiler and systematizer from Greek to Latin. In addition the work of Leucippus (500–440 BC) should be considered a precursor to atomic theory but only a fragment of his work survives, including quotations appearing in the work of other authors.

After this brief digression about materialism in general let us go back to historical materialism itself, meaning the history of Marxist theory and the materialist conception of history. If we look at Engels' introduction for *From Utopian Socialism to Scientific Socialism* (Engels 1962) one reads: *Historical Materialism designates a particular vision of history development which looks for a final cause and the driving force of all important historical facts in the economic development of society, in the transformations of the modes of production and exchange, in the division of society in different classes and finally in classes struggle.*

According to Marxism, historical development depends on how society organizes itself economically and is determined by means of these economical transformations including those resulting from class struggle. Engels considered Marx as the author of Historical Materialist theory and believed that this theory, along with surplus-value theory, constituted Marx's primary achievements. Obviously

[4] Prigogine (1996) refers to Epicurus and says: "It was Epicurus the first to establish the terms of the dilemma which modern physics gave the weight of its authority. He was the successor of Democritus and thought the world constituted by atoms in motion in empty space. He also thought that atoms fall down with the same velocity according to parallel trajectories. How could they collide? How could appear novelty, a new combination of atoms? To Epicurus, science problem, nature intelligibility and the destiny of men were inseparable. What meant human freedom in the deterministic world of atoms?"

[5] It was Lucretius who introduced the idea of *clinamen* in Epicurus´s atomic theory. This word means declination and poses the possibility of change in the deterministic framework of Epicurus´s atomic theory. It is also an open question as to whether Epicurus may have mentioned such a term in a piece or fragment of writing that got lost (Paraire 1997).

Engels minimizes his own contribution, but because their common work began with *German Ideology* around 1845–1846 (Marx and Engels 1980) or possibly even some years before, the contributions of both men cannot be separated. Some elements of Historical Materialism do exist in previous Marx–Engels writings. From a scientific point of view we can say that Historical Materialism has an empirical basis or that its theory is a collection of empirical theses that Marx and Engels were progressively building, the scientific character of which depends on its potential for explanation and also its predictive capacity. It is important to remark that predictability does not mean a kind of economic determinism despite several "Marxists" having adopted this type of determinism-mechanicism.

Marx, in the preface to his *Contribution to Critique of Political Economy*, published in 1859 (1971), described Historical Materialism as follows:

In the social production of their life, men enter into definite relations that are indispensable and independent of their will, relations of production which correspond to a definite stage of development of their material productive forces. The sum total of these relations of production constitutes the economic structure of society, the real basis, on which rises a legal and political superstructure, and to which correspond definite forms of social consciousness. The mode of production of material life conditions the social, political and intellectual life process in general. It is not the consciousness of man that determines their being, but, on the contrary, their social being that determines their consciousness. At a certain stage of their development, the material productive forces of society come in conflict with the existing relations of production, or—what is but a legal expression for the same thing—with the property relations within which they have been at work hitherto. From forms of development of the productive forces these relations turn into their fetters. Then begins an epoch of social revolution. With the change of the economic foundation the entire immense superstructure is more or less rapidly transformed. In considering such transformations, a distinction should always be made between the material transformation of the economic conditions of production, which can be determined with the precision of natural science, and the legal, political, religious, aesthetic or philosophic—in short, ideological forms in which men become conscious of this conflict and fight it out. Just as our opinion of an individual is not based on what he thinks of himself, so can we not judge of such a period of transformation by its own consciousness; on the contrary, this consciousness must be explained rather from the contradictions of material life, from the existing conflict between the social productive forces and the relations of production. No social formation ever perishes before all the productive forces for which there is room in it have developed; and new, higher relations of production never appear before the material conditions of their existence have matured in the womb of the old society itself. Therefore mankind always sets itself only such tasks as it can solve; since, looking at the matter more closely, it will always be found that the task itself arises only when the material conditions for its solution already exist or are at least in the process of formation. In broad outlines Asiatic, ancient, feudal, and modern bourgeois modes of production can be designated as progressive epochs in the economic formation of society. The bourgeois relations of production are the last antagonistic form of the social process of production—antagonistic not in the sense of

individual antagonism, but of one arising from the social conditions of life of the individuals; at the same time the productive forces developing in the womb of bourgeois society create the material conditions for the solution of that antagonism. This social formation brings, therefore, the prehistory of human society to close.[6]
In the same preface Marx also mentioned a previous independent investigation in the same direction by Engels and emphasized the importance of writings such as the Communist Manifesto (Marx and Engels 1998).

The long Marx quotation cited above is referred to many times by several authors that studied Marx and Engels' Historical Materialism. It is curious that Marx does not mention the question of class struggle. Finally it is worth mentioning that the question of how to pass from one mode of production to another, the old question discussed by the revolutionary movement in 1917 in Russia, remains an unsolved problem in Marxism to this day.

1.1.3 Dialectical Materialism

The creators of dialectical materialism were keenly conscious of the gap in their theoretical work in terms of nature sciences. Marx looked forward to writing a handbook of dialectics after he had completed *The Capital*. In the latter years of his life, Engels made a comprehensive study of mathematics and natural sciences in order to reconstruct their theoretical foundations with the aid of the materialist dialectics, just as he and Marx had previously revolutionized social sciences. Engels wrote in the second preface to Anti-Dühring (Engels 1971):

Marx and I were pretty well the only people to rescue conscious dialectics from German idealist philosophy and apply it in the materialist conception of nature and history. But knowledge of mathematics and natural science is essential to a conception of nature which is dialectical and at the same time materialist. Marx was well versed in mathematics, but we could only partially, intermittently and sporadically keep up with the natural sciences. For this reason, when I retired from business and transferred my home to London, thus enabling myself to give the necessary to it, I went through as complete as possible a "moulting", as Liebig calls it, in mathematics and the natural sciences, and spent the best part of 8 years on it.
Anti-Dühring was the first fruit of this work and *Dialectics of Nature* the last. While the first book remains the best exposition of the philosophy of dialectical materialism, the second, despite its fragmentary character, must now be read as its indispensable supplement. *Dialectics of Nature* could not be finished because of the tremendous effort employed to edit and publish *The Capital*. This is also concrete proof of the advancement of natural science by social scientific development (Althusser 1967; Fataliev 1966).

---

[6] For a more systematic and modern explanation of Historical Materialism, see (Cohen 2000).

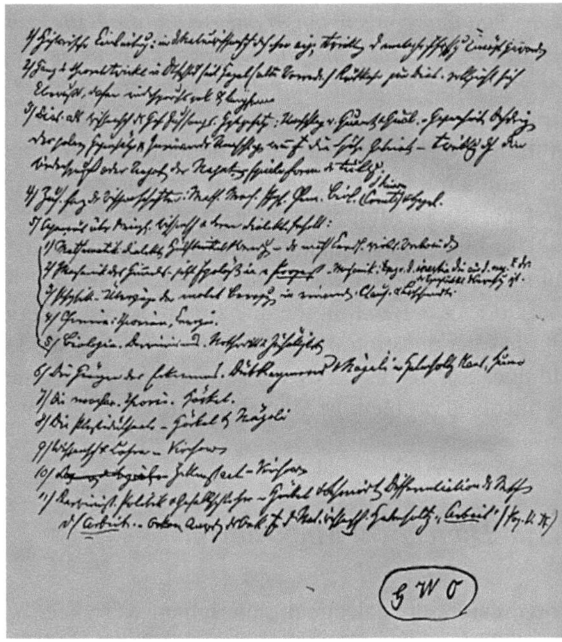

Fig. 1.3 Engels's outline of Dialectics of Nature

In *Dialectics of Nature* (Fig. 1.3) Engels aimed to demonstrate that the process of nature obeys the same general laws of motion as social and intellectual processes. These dialectical laws operate similar to those which have historically governed the apparent fortuitousness of events; *the same laws as those which similarly form the thread running through the history of the development of humanity and gradually rise to consciousness in the mind of man.*

The central idea of *Dialectics of Nature* is the classification of the forms of motions of matter as well as the classification of the sciences which treat these forms. A pure change of place is the lowest form of motion, while thought is the highest. The mechanical, physical, chemical and biological forms of motion are the main forms dealt with by the natural sciences. Each lower form of motion changes into a higher one by a dialectical process. Each higher form of motion contains, but does not come down to, a lower form as a subordinate element. On the basis of this theory of the forms of motion of matter, Engels built the dialectical materialist classification of the natural sciences, each of which *analyses a single form of motion, or a series of forms of motion that belong together and pass into one matter.*[7]

In the *Introduction,* Engels presents a critical review of the development of natural science on its theoretical side. He explains how and why the viewpoint of the

[7] Recently, some authors have written about the dialectics of nature based on new knowledge about biological processes (Prigogine 1996).

absolute immutability of nature was dominant in the first period of natural knowledge. Throughout the text Engels examines the dialectical content of mathematics, mechanics, physics, chemistry and biology. In mathematics he singles out the problem of the seeming apriority of mathematical abstractions, in astronomy the problem of the origin and development of the solar system, in physics the theory of the transformation of energy, in chemistry the problem of atomistic, in biology the problem of the origin and essence of life, the cell theory, and Darwinian theory of evolution.

Dialectics of Nature deals more extensively than any other work written by Marx and Engels with such problems and categories of dialectics as causality, necessity and chance, classification of the forms of judgment, the relationship of induction and deduction, and the role of hypothesis as a form of development of natural science.

Engels in the chapter dedicated to dialectics enunciates the three laws of dialectics as follows: "It is, therefore, from the history of nature and human society that the laws of dialectics are abstracted. For they are nothing but the most general laws of these two aspects of historical development, as well as of thought itself. And indeed they can be reduced in the main to three: the law of the transformation of quantity into quality and vice versa; the law of the interpenetration of opposites; the law of the negation of the negation."[8]

The first law signifies that: "in nature, in a manner exactly fixed for each individual case, qualitative changes can only occur by the quantitative addition or subtraction of matter or motion (so-called energy)". This law is observable in chemistry where the properties of bodies are altered in concordance with their changed quantitative composition. Another clear example is the periodic arrangement of the elements according to their atomic weights.

The second law asserts that everything has a self-contradictory character, containing within itself its own opposite. Engels studies in the third chapter the most important of scientific problems, the basic forms of motion. All natural knowledge is based in the conception of material movement of one kind or another. A correct conception of motion is therefore absolutely indispensable to natural science. Motion to Engels is a contradictory combination of attraction and repulsion. Every mode of motion in nature from the lowest to the highest, from simple mechanical motion to complicated organic behavior, embraces and arises out the simultaneous action and reaction of attraction and repulsion.[9]

The third law of the negation of the negation, which Hegel used as the fundamental law for the construction of his whole system of thought, has a far wider sphere of application in the system of nature. This law expresses the fundamental form of development in nature. The whole process initiates with the first negation that in turn undergoes self-differentiation and division until it, too, passes into

---

[8] The three laws cited above appear in the first chapter of Engels´s *Dialectics of Nature*, p. 62 (1966).

[9] Engels's conception of motion is similar to the theoretical framework proposed by the Laplacian project, as we will see later.

its own opposite and thereby becomes negated. The final result of this process is called the negation of the negation, a synthetic unit which has discarded the transitional forms but preserved within itself the essential content of both sides of the contradictory whole.

Historical Materialism means knowledge of the laws of human evolution, or partial knowledge that can be elevated to a theory of superior degree by an integration process accomplished by Dialectical Materialism. Historical Materialism is a human science and Dialectical Materialism is a science of global reality (Lefebvre 1940).

1.1.4 Theory of Knowledge

Despite Marx having not written anything about dialectic, he used this method extensively in his political and economic writings. Only Engels, mainly in: *From Utopian Socialism to Scientific Socialism, Ludwig Feuerbach and the end of German Classical Philosophy, Dialectics of Nature and Anti-Dühring*, provided a general overview of a Marxist theory of knowledge. But it is quite obvious that Engels had discussed dialectics with Marx to the extent that this method was a common theme in the work of both founders of Marxism. Other famous Marxists have contributed to the development of this theory. Lenin published *Materialism and Empirocriticism* in 1908. Georgy Luckacs (1885–1971) and his disciple Gyorgy Markus (1934), both Hungarian philosophers, also wrote on this subject (Lukács 1971; Markus 1974).

Our main objective in this section is to present some important characteristics of the Marxist theory of knowledge in order to understand the debates and controversies within Marxism. In addition we will systematize from a didactic purpose the main characteristics of such a theory. First of all Marxist theory emphasizes in a special way the question of objectivity, meaning the independence of reality from our ideas or other kind of representations of reality. In other words, to Marxist theory reality has an ontological character.

A second question is the role of the labour process for acquiring knowledge and therefore the social character of knowledge. These two fundamental characteristics, objectivity and labour, have discarded empiricism, idealism, skepticism and dogmatism as isolated forms that cannot alone explain the knowledge process. For a deeper understanding of these two questions, see Marx's thesis on Feuerbach (Labica 1990). In this work Marx criticizes both materialism and idealism. Marx refers to Feuerbach's materialism and analyses knowledge as knowledge about the world created by man that does not exist outside history, society and industry and which is not taken into account by Feuerbach. With respect to philosophical idealism Marx shows that this philosophy has developed the active aspect of reality that is the same as subjective activity within the knowledge process. The knowledge process is not the perception of objects from immediate reality but a consequence of human activity. In this context Marx recognizes the importance of idealism in the relation subject-object. On the other hand this point of view only considers consciousness and its abstract character, which does not include practical activity as a fundamental aspect of reality.

Taking into account previous remarks as well as the questions discussed we can try to systematize some fundamental aspects of a Marxist theory of knowledge as follows:

(a) As Lenin stated, Marxism is a new form of materialism. Consequently, to Marxism reality has its own existence independent of our mind or our ideas as well as any kind of mental representation (1971).
(b) Marxism implies that being has precedent over thought or that existence precedes consciousness. This statement does not mean that Marxism neglects the existence of spiritual qualities and other attributes of sensibility. On the contrary. Marxism affirms that these qualities spring from social being and depend on our material existence.
(c) Marxism regards reality in permanent motion similar to certain pre-Socratic Greek philosophers. This statement means that both social and natural reality can be seen as processes with a beginning, development and death. As a corollary of this statement, to apprehend reality it is necessary to adopt a perspective of motion that understands all moments of the process and identifies its multiple connections and relations.[10]
(d) In spite of an absence of consensus among Marxists, Engels' writings consider dialectics as belonging to the structure of reality and not only as a method for analyzing reality. Otherwise it is complete nonsense.[11]
(e) We are considering reality as the social world, nature and thought. Dialectics operates on these three dimensions as a unifying principle and in spite of the specificities of each one from a scientific point of view, which implies particular sciences to study each individual field of knowledge, they belong to the same reality. In addition there is no consensus among Marxists with respect to the existence of dialectics in natural processes. These disagreements in some cases affirm that the above point of view represents a metaphysical vision (Lukács 1979). Recent developments in system dynamics where the concepts of complexity and chaos appear have encouraged new studies on the dialectics of nature. These new concepts show its applicability to biology, cosmology and the origins of life.
(f) The superstructure of society, the political, religious and legal ideas are determined and conditioned by economic structure. This statement is equivalent to

[10] Engels refers to the limitations of the old mechanicist-materialism by saying: "The second specific limitation of this materialism was its incapacity to conceive the world as process, as something obeying historical development. This fact had a correspondence with the stage of development of the nature forces of that period, a metaphysical manner, an anti-dialectical philosophy related to that".

[11] In the first chapter of (Engels 1966), Engels states: "Dialectical laws, although, are obtained from the history of nature as well as the history of human society. These laws are the more general laws from both phases of the historical development with the human thought".

saying that all forms of thought depend on how men relate to each other in economic production as well as in the circulation and distribution of goods. It is important to explain that this kind of dependence does not mean determinism of the Newtonian type. In a letter addressed to Bloch, Engels postulates that in some cases other non-economic causes can be fundamental to social transformation.[12] The dependence of superstructure on the material basis is a general indication and not a general law applied to any situation.

(g) As a corollary of item (b), the mode of production establishes limits and new horizons to thought. This is the same as to say that man is limited by his time. Engels considers that modes of production are different in space and time. In addition neither Marx nor Engels has stated that there is a sequence of social development towards a developed situation: primitive communism, slave society, feudal society, capitalism and socialism or communism (Engels 1971).

(h) To Marxism class struggle is a pivotal point, the contradiction that serves as the historical motor of the social world.[13] It is worth mentioning that the concept of class struggle is not a Marxist concept but was adopted by French Historians from the Restoration period. Besides this conception, there are several historians and politicians that believe in another kind of historical motor than the development of productive forces. They are generally known as economicists. Another important consideration is that Marx and Engels gave an exaggerated importance to the development of productive forces as a kind of inexhaustible source of societal progress. They were probably mistaken, although they did criticize the soil degradation and other environmental problems that are nowadays the concern of ecologists.

(i) From a stricter point of view, Marxism disagrees with the dichotomy that knowledge must come either from experience or sensation or from reason. To the Marxist, both sources must be incorporated in the theory of knowledge. Empirical results as well as other information from reality should be articulated and connected by reason, and by doing this there is no estranging addition to the real apprehension. Reciprocally, reality can be known by man since historically he could build the necessary tools for this achievement. In this sense the real is cognoscible by man in contradiction to Kant's statement.[14]

---

[12] See the letter addressed by Engels to J. Bloch written in London on 21–22 September of 1890 (Marx and Engels 1965).

[13] In the Communist Manifesto, Marx and Engels affirm: *The history of every society until now is the history of classes struggles* (Marx and Engels 1998).

[14] Piaget (1997), p. 247, discusses in depth the problem of innate ideas. He said: "There are no "innate ideas" in the Cartesian sense, etc. It is true that by generalizing we could consider that as the categories of "a priori" in Kant´s sense. In epistemology, the existence of "a priori" synthetic judgments was assumed by Henri Poincaré related to number intuition (in the sense of iteration $n + 1$) and to "group of displacements" … Although, from the psychogenetic point of view, such interpretations do not resist to examination…".

1.2 Genetic Epistemology

Jean Piaget (1896–1980) in a series of lectures at Columbia University, published by Columbia University Press in 1968, defines genetic epistemology as *an attempt to explain knowledge, and in particular scientific knowledge, on the basis of its history, its sociogenesis, and especially the psychological origins of the notions and operations upon which it is based. These notions and operations are drawn in large part from common sense, so that their origins can shed light on their significance as knowledge of a somewhat high level. But genetic epistemology also takes into account, wherever possible, formalization—in particular, logical formalizations applied to equilibrated thought structures and in certain cases to transformations from one level to another in the development of thought.*

As we can see from the above definition, genetic epistemology as a scientific discipline was established by Piaget. Its goal is to link the validity of knowledge to the model of its construction, thus affirming that the method in which the knowledge was obtained or created affects the validity of that knowledge. As an example, our direct experience with gravity makes our knowledge of it more valid than our indirect experience with black holes. Genetic epistemology also explains the process of how a human being develops cognitively from birth throughout his life.[15]

Over a period of six decades, Jean Piaget conducted a program of naturalistic research that has profoundly affected our understanding of the acquisition of knowledge, as well as child development. He called his theory Genetic Epistemology because he was primarily interested in how knowledge developed in human organisms. Piaget used concepts from biology and philosophy and these disciplines influenced his theories and research of child development.[16]

To Piaget the knowledge process has some general characteristics:

- Knowledge has a biological function, and arises "tied to action", as Piaget said.
- Knowledge is basically *operative*, meaning that it is about change and transformation.
- Knowledge consists of cognitive structures.
- Development proceeds by the assimilation of the environment to these structures, and the accommodation of these structures to the environment.
- Movement to higher levels of development depends on *reflecting abstraction*, which means coming to know the properties of one's own actions, or coming to know the ways in which they are coordinated.

[15] Piaget is the first to conceive and to construct a bridge between speculation and an experimental science of the theory of knowledge. His investigations created an appropriated basis to gnoseology. In addition, it was possible to formulate epistemological problems in a language that permits the establishment of experimental control and thus specific ways to guide experimentation.

[16] In order to develop an experimental basis to improve the understanding of human knowledge, Piaget has given important contributions to particular sciences, especially to child psychology, as a fruitful way to study intelligence psychology.

The concept of cognitive structure has a central importance in Piaget's theory. These structures are patterns of physical or mental action that underlie specific acts of intelligence and correspond to stages of child development. There are four primary structures or stages of development. Progress from one stage to another comes from an assimilation, which occurs when the perception of a new event or object appears to the learner in an existing schema and is usually used in the context of self-motivation. Another moment of the cognitive process is an accommodation, meaning that experiences are accommodated according to the outcome of the tasks. The highest form of development in the whole process is equilibration which encompasses both assimilation and accommodation as the learner changes their way of thinking in order to arrive at a correct or different answer.

According to Piaget, human development is divided into four major stages, or periods. Other stages beyond the forth were also considered possible and each stage can be divided into sub-stages. The four stages are: (a) we begin at birth with the *sensoriomotor* stage.[17] During this stage we are limited to *thinking in action*. Many accomplishments have taken place by the end of this period as a full understanding of permanent objects and the ability to imitate someone else's action on the basis of memory alone; (b) around 2 years of age, we enter into the *preoperational* stage. This stage is marked by the ability to anticipate, that is, to consider possibilities before acting on them. It is also characterized by the appearance of the semiotic function, which embraces speaking and understanding language. However, many elementary forms of logical reasoning are not yet available; (c) around 6 or 7 years of age, we achieve concrete operations. During this stage, logical structures for hierarchical classification become available, as do structures for seriation, conservation of physical quantities and mathematical operation of numbers; (d) from age 12, we move into formal operations. Formal thinking is marked by the ability to generate and work with larger spaces of possibilities systematically, including possibilities that are quite abstract.

Two correlated books deal with the development of knowledge. The first is *Biology and Knowledge* (Piaget 1997), in which Piaget discusses problems of intelligence and knowledge in general and as logical mathematical problems using concepts of modern biology in particular. In *Psychogenesis and History of Sciences*, published by Piaget in collaboration with Rolando Garcia in 1980, a few months before his death, the two authors tried to establish relations between the development of knowledge and the history of sciences. This book is an attempt to determine if mechanisms of evolution from a historical period to the subsequent contexts of notional systems such as algebraic, geometric and mechanics maintain an analogy with changes in genetic stages as mentioned above (Table 1.1).

[17] Piaget postulated that the starting point of knowledge is not sensation but sensoriomotor activity. This concurs with Marx's perspective emphasizing activity, the *praxis*, as the fundamental aspect of knowledge.

Table 1.1 Relation between stages of knowledge and logical systems

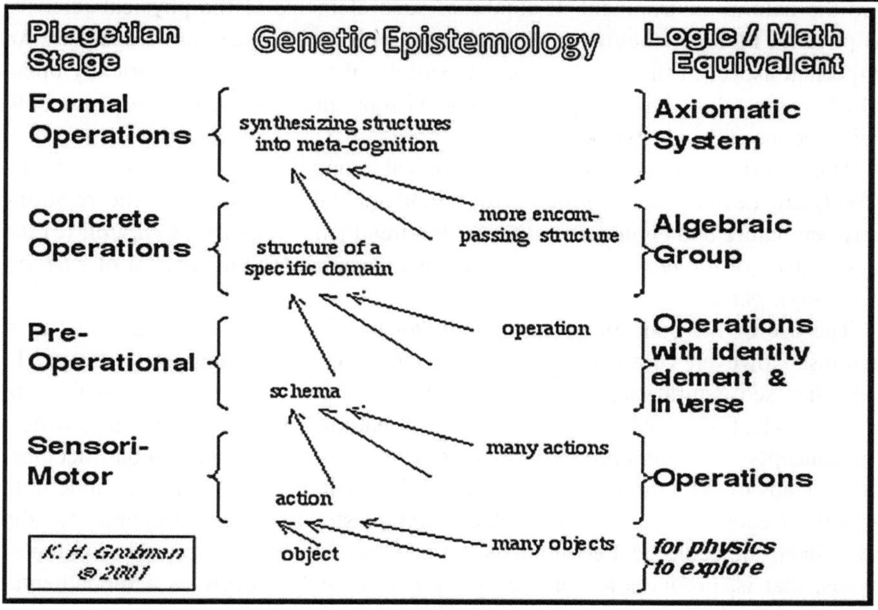

Piaget also proposed three types of knowledge: physical, logical mathematical and social. Physical knowledge refers to knowledge related to objects in the world, which can be acquired through perceptive properties. Logical mathematical knowledge is abstract and must be invented but through actions on objects. Finally, social knowledge is culture specific and can be learned from other people within one's cultural group.

1.3 Ecological Economics

Ecology describes nature as consisting of complex systems with many parts and reciprocal functions. Among these are biological communities, which are interacting groups of organisms, and ecosystems, which are biological communities together with their nonliving environments. The term ecology (*ökologie*) was coined by the German biologist Ernst Haeckel (1834–1919) in his *Generelle Morphologie der Organismen*, published in 1866. He also defined the new science as follows:

> By ecology we mean the body of knowledge concerning the economy of nature—the investigation of the total relations of the animal both to its organic and inorganic environment; including above all, its friendly and inimical relations with those animals and plants with which it comes directly or indirectly into contact—in a word, ecology is the study of all those complex interrelations referred to by Darwin as the conditions of the struggle for existence.

This new discipline, called ecological economics, is a multidisciplinary branch of knowledge which has by subject the interconnections between economic systems and the natural environment. It combines such elements of the physical sciences as physics, biology, chemistry, geology, etc. with tools of economic analysis. An important aspect of these studies is that ecological economics is historically open, which means that it is receptive to new visions and new approaches into both fields: economic and environment.

The contributions of Marxism to ecological economics as proposed by Burkett (2009) are developed in terms of four fundamental questions: (a) the relations between nature and *economic value*; (b) the treatment of *nature as capital*; (c) the significance of the *entropy law* for economic systems; (d) the concept of *sustainable development*.

The use of the concept of *natural capital* has its origins in neoclassical economics. However, ecological economists have tried to popularize its usage. In fact, they see natural capital as a useful metaphor for legitimizing ideas that are pro-ecological. Obviously different views on natural capital correspond to different concepts of *sustainable development*. The dominant thought among neoclassicists leads to the notion of *weak sustainability*, which means continued indefinite growth of economy even with the decreasing of stocks of natural capital. On the other hand the ecological economists see natural and produced capital as complements, that is, production constrained by *strong sustainability* as a fundamental condition which implies that natural capital be non-decreasing.

In the second question about nature as capital, Paul Burkett postulates: "Marxism deepens the critique of monetary and market valuation by rooting it more firmly in the relations of production. In the Marxist view, not only are generalized market exchange and monetary valuation a function of the separation of workers from necessary conditions of production, but the same separation enables market prices to be regulated by the abstract; (homogenous) labour time objectified in commodities. Burkett continues: While recognizing that workers often struggle in ways that do not fundamentally question wage-labour and the capitalization of nature, Marxism detects a radical potential for worker-community movements to fight for new relations of production that treat human-natural relations as ends in themselves rather than instruments of alienated production and profit-making".

With respect to the third question, the relation between the entropy law and economic systems, we will try to present some important aspects of this debate mainly from a Marxist perspective. As we know, entropy is a measure of the total disorder, randomness, or chaos in a given system, which means increased entropy implies greater disorder. The second law of thermodynamics prescribes that entropy of an isolated system is non-decreasing. In other words, energy is only transformed from a more ordered stage to less ordered one. Another view of this situation is that heat will not flow spontaneously from a cold to a hot object in an isolated system. From a human standpoint usefulness of materials must obey the second law of thermodynamics which implies that all kinds of energy transformation convert energy into less available and less useful forms.

If we look at the economic process, the importance of the entropy law was first proposed by Nicholas Georgescu-Roengen (1906–1994) and Herman Edward Daly (1938). The considerations made by these two authors were initially very simple. They observed that production depends upon materials and energy that are provided by nature. The economic process only absorbs matter-energy and throws it out continuously. They then argue that matter-energy comes in the economic process in a form of low entropy and leaves it in a high entropy stage. Thus, the economic process combines human labour with low-entropy materials transforming them and energy from more ordered forms into less ordered forms.

Georgescu, after explaining entropy applied to economy with several examples, states: "There are several lessons to be derived from this analysis. The first lesson is that man's economic struggle centers on environmental low entropy. Second, environmental low entropy is scarce in a different sense than Ricardian land… the entropy law is the reason why an engine (even a biological organism) ultimately wears out and must be replaced by a new one, which means an additional tapping of environmental low entropy". This citation and the following were extracted of Georgescu's book: *Energy and Economic Myths* (1976).

Georgescu discusses the fundamental questions of sustainability as follows: Actually, the problem of the economic use of the terrestrial stock of low entropy is not limited to the mechanization of agricultural only: it is the main problem for the fate of the human species. To see this, let S denote the present stock of terrestrial low entropy and let r be some average annual amount of depletion. If we abstract (as we can safely do here) from the slow degradation of S, the theoretical maximum number of years until the complete exhaustion of that stock is S/r. This is also the number of years until the industrial phase in the evolution of mankind will forcibly come to the end.[18]

The fourth question to be discussed is the concept of *sustainable development.* Burkett (2009) quotes the Brundtland Report of the World Commission on Environment and Development, published in 1987: *Sustainable development is development that meets the needs of the present without compromising the ability of future generations to meet their own needs.* This concept becomes the dominant concept in the study of interactions between economy and the biophysical environment. Independent of the different visions, the fundamental problem remains as to how natural limits could, in absence of appropriate adjustments in our resources, make development unsustainable.

[18] Georgescu offered eight concrete recommendations to move human society in the right direction: (1) the complete prohibition of weapons production, thereby releasing productive forces for more constructive purposes; (2) immediate aid to undeveloped nations; (3) gradual decrease in population to a level that could be maintained only by organic agriculture; (4) avoidance, and strict regulation if necessary, of wasteful energy use; (5) abandon our attachment to *extravagant gadgetry*; (6) *get rid of fashion*; (7) make goods more durable and repairable; (8) cure ourselves of workaholic habits by rebalancing the time spent on work and leisure, a shift that will become incumbent as the effects of the other changes make themselves known.

Social movements work with two fundamental concepts: climatic and ecological debts. Climatic debt is based in the situation of the present relationship between northern and southern countries. Northern countries, obviously industrialized and well developed, grew up based on unlimited consumption of combustible fossils being a fundamental cause of climatic crisis. In addition, these countries have appropriated the atmosphere and the oceans as a "global common good", as well as modifying the capacity of carbon absorption of the biosphere. Another important consideration is that these countries represent less than 20 % of the world population and yet their atmospheric emissions are around 75 %. The climate debt presupposes that is possible to share the obligations and responsibilities in a just and equilibrated way in order to repair and reduce carbon emissions. The second concept, the ecological debt, contributes to a different analysis of international relationships and the interchange between north and south beyond economic terms. It works to construct tools to repair environmental damage and to guarantee reparation from and punishment of those responsible. Consequently, new arguments are presented in order to cancel financial debts accumulated in an illegitimate manner.

Ecological economics, with diverse visions about sustainability, has the basic concern of natural limits to economic development. On the other hand a fact persists in the majority of analysis in the overexploitation of natural resources. It reduces their availability now and in the future. Conversely, sustainable exploitation requires a specific definition depending on the particular resources in question.

1.4 Theory of History and General Historiography

Lucien Goldman (1913–1970) (Goldmann 1970), postulates that *all social fact is historical with the converse also being true.* In other words, history and sociology *study the same phenomena.* The solution in Goldman's vision is not to collect results from sociology as well as history, but to abandon the whole abstracts of sociology and history in order to achieve a concrete science of human facts. This can be synthesized in a historical sociology or, by the same measure, a sociological history. He proposes:

What we are looking for in the past knowledge is the same that we are looking for in the contemporizing men knowledge. First of all the fundamental attitudes of individuals and human groups facing the values, the community and the universe. If, historical knowledge shows us a practical importance, is because in that we learn to know men that, in conditions and, using different ways, often not applied to our epoch, they fighted for values and similar ideals, identical or opposed to that we have today; what give us consciousness of pertaining to a whole that transcend us, and that nowadays we gave continuity, is that men that come after us will continue in the future. Historical consciousness does exist even for an attitude that overcomes the individualist being, and it is precisely one of principal means to accomplish this overcoming. To rationalism, the past is an error whose knowledge is useful to eliminate the progress of rationality; to empiricism the past is a mass of real facts, that are exact with respect to a conjectural future; only a dialectical attitude can accomplish the synthesis, encompassing the past as a stage and necessary way, worth to a common action of men of the same class of the present, in order to build an authentic community and universal in the future (Goldmann 1970).

It seems clear from the long quotation above that the object of historical sciences is built by human actions from all places and all times. Thus, history's goal, from Goldman's viewpoint, encompasses collective movements such as revolutions and wars, as well as the individual acts of people like Napoleon and other important historical figures. Goldman also made a critique of the rationalist and empiricist movements. In addition he states the importance of adopting a dialectical approach in order to apprehend history as a process in continuous development.

About the important question of scientism and determinism raised by some Marxists, Goldman wrote: "the fundamental epistemological error of this scientism is to consider human community as object of study. Object, in this case should be understood as something extern and outside". Obviously, this viewpoint about historical fact separating being and object belongs to a history of the positivistic view.

With respect to the economicism of certain Marxists, Goldman wrote: "there is in men's life not a unique and decisive importance but a kind of privilege associated to economic causes? In fact a history that has developed until now, the answer is yes. This happens because man and society are global facts where is impossible to separate lays. Man is an alive and consciousness being belonging to world with many realities as economical, social, political intellectual and religious. He faces global action of this world and reacts to it. It is a kind of dialectical relation. As result individual consciousness is influenced by all these factors and maintains its unity and coherence".

This is enough to explain and confirm the privilege of economic factors action in past and contemporizing history because men are made such that, to love, to think or to believe, they must live, they must eat and use cloths. These domains of human activity can have less importance to thinking and other activities with the condition that the satisfaction of all necessities be completely supported and thus men can dedicate to them a small fraction of total activity. What is important is that people, dominated classes or primitive society's members always lived under privation and they were obliged to spend in labour a lot of time and—in modern world—they became insecure and permanently scared with respect to future (Goldmann 1970).

Goldman's words, mainly at the end of the citation, come across as prophetic speech. Obviously the influence of economic factors in history is not a mere invention of Marxists and not a matter of opinion but the product of historical evolution. Men who live to produce, etc. are subject to relations of this nature and, in addition, these influences are different to different social classes or groups. When these conditions are modified, history itself will be influenced by other factors and other constraints. To Marxism there is no dictatorship of economic factors or something intrinsic in the history of economic nature and the extra-historic.

Before studying historiography itself, let us look at a brief description of the question of ideology in history according to a Marxist vision.[19] If we look at nature sciences like physics, geology, biology, etc., the influence of such subjective factors as opinions, judgments, and ideological questions are different than within the social or human sciences. In other words the relationship between being and object in the field of nature sciences is fundamentally different than that within the human sciences. In the second case being and object coincide practically. It is also important to note that even nature sciences are affected by ideological questions, but in a different way, because all manner of science is socially produced, being influenced by the social interests and political structures that govern the world.

In human sciences the problem is more delicate because of the inseparability of these sciences and ideological influences. The central question to be analyzed is how to consider scientific knowledge in this field. How to establish a difference between *doxa* and *episteme* in a Greek sense?

Apparently human sciences are subject to an avoidable relativism. The question that we can discuss is: if it is impossible to separate subjective factors and moral judgments from social facts, what opinion should prevail? Are all opinions equivalent or can some of them be disregarded? Marxism tries to escape from this relativist trap without adopting a positivist viewpoint, postulating that there is a privileged point of reference for regarding historical facts, namely, the viewpoint of oppressed and exploited classes. Conversely, politically and economically dominant classes all have an interest in hiding and mystifying historical truth. It being impossible to separate ideological aspects from the facts of social life, we can regard social facts from the viewpoint of exploited classes. To Marxism this is the best approach to historical knowledge.[20] Two other possibilities are something similar to Galilean relativism, which considers all systems of reference with constant velocities equivalent, and the positivist option, which proposes to separate historical fact from moral judgments.

After this brief digression about historical method and certain difficulties for historians, let us study historiography, which should be understood as the study of historical writings, its methods, interpretations and controversies. As an academic discipline, historiography is a comparatively recent autonomous branch of knowledge. It began in the 20th century. However, the basic concern of historians for comparing their work with that of their predecessors is very old (da Silva 2000; Tétard 2000).

The founders of the historical sciences are Herodotus (484–425 BC)(1988) and Thucydides (460?–400? BC) (Tucídides 1999). The former was born in

[19] See Löwy in (1988). On p. 11 one can read: *The concept of ideology was not coined by Marx: he only has taken it again. He was literally invented (in the true sense of invent, to take off from nothing) by a French philosopher few known, Destutt de Tracy, disciple of less category from encyclopedists, who published in 1801 a book called "Elements of ideology".*

[20] For a discussion of this question in depth, see introduction (Löwy 1987), p. 195 and following.

Halicarnassus, Caria (now Bodrum in Turkey), Asia Minor. He was exiled from his native land for political reasons and subsequently became a great voyager. He was the first to collect materials systematically, testing their accuracy and arranging them in a well-written narrative. His masterpiece, *The Histories*, is a record of his *inquiry*, a word that gained and acquired a modern meaning as *history*, into the origins of the Greco-Persian wars. It was divided by later editors into nine books named after the Muses. His content traces the growth of the Persian Empire beginning with Croesus of Lydia, though Cyrus and Xerxes. The most important event in *The Histories* is the Battle of Marathon (490 BC) where Persians were defeated by the Greeks.

The title of Father of History is granted to Herodotus because, before him, historical documents were made by the so-called logographs, or writers of prose, in contrast with writers of verse, who could be poets or philosophers, and in certain historical periods used poetic form to express themselves. In addition, Herodotus is one of the primary sources of information on ancient lands and people with important anthropological and geographical details.

Thucydides was an Athenian aristocrat and a historian from the period (460–395) BC His *History of the Peloponnesian War*, a conflict with a duration of 27 years (431–404), details the war between Sparta and Athens. He died 21 years into the war in 411/410 B. C, and this the work was interrupted.

He is also known as the father of *scientific history* because his narrative is, in fact, an analysis in terms of cause and effect without reference to intervention by gods. Another characteristic of his work is a political realism because the relations between nations are based on might rather than right. This last aspect of his text is studied at advanced military schools.

Thucydides, living in the Athens of the Pericles period, regarded the reasons of statesman and actions of government as the essence of history.

As we have named Herodotus the father of history, Polybius (208? – 125? BC) (1996) can be considered the *father of historiography.* He was born in Arcadia around 200 BC Through study of his precursors, he became conscious of the importance of the subject of Rome's supremacy over Greece at the time that he lived. His work offers a panoramic view and a critique of texts about Roman history written by those precursors. Polybius also proposed a *general history* instead of a *partial history.*

Polybius wrote several works, the majority of which are lost. He is considered by some historians as successor of Thucydides with respect to objectivity and critical thought.

Another important ancient Greek historian is Diodorus Siculus who lived in the period (90 – 20 BC). He was born at Agyrium in Sicily, now called Agira. His work consisted of 40 books of which 1–5 and 11–20 survive. Part of these books addresses the history of the non-Hellenic and Hellenic tribes up to the destruction of Troy. He also describes the history and culture of ancient Egypt, Mesopotamia, India and others. The importance of his work lies in the bridge he built between Greek civilization and the emergent Roman society. He named his universal history the Historical Library, accomplished in the period 60 – 30 BC Like Polybius, he critiqued his precursours, especially those who worked in Hellenistic historiography.

From the viewpoint of a critique of historiography, Lucian of Samostata (125? – 192?) is an important name. He claims to have been born in Samostata, a former kingdom of Commagene that had been absorbed by the Roman Empire and made part of Syria. He wrote a work entitled *How to Write History*, written in the year 165. This text is a critique of the formal aspects and historiographical content of his contemporaries. He considers the objective of historians to be to tell *what happened* instead of wasting time with anything other than objective truth. Hence, he anticipates by the purposes of modern historical schools by some centuries.

During the Roman civilization, historiography underwent considerable development. Unlike the Greek civilization, the Roman state had great concern with preservation of historical records from its earliest days. The most famous names in Roman historiography are: Julius Cesar (100–44 BC), Sallustius (86–35 BC), Titus Livius (59 BC–17), Tacitus (55?–117) and Suetonius (75?–150).

From the middle ages, we can find several important historians: Jordanes in the 6th century, author of a *History of Goths* (551), Gregory of Tours (538–594), author of *History of Francs*, Isidore of Seville (570–599), author of *History of Goths, Vandals and Swabians*, Beda (673–735), author of *Ecclesiastic History of Englishmen*, and Paul Diaconus (720–799), author of *History of Lombards*. In these historical writings a *sacred history* and an *unholy history* are inseparable.

At the end of Middle Ages profound social transformations occured in Western Europe, mainly at the beginning of the 11th century, with accelerated trade development and the growth of both urban centers and an urban culture and tradition as a result of this process. The impact of this new social framework on historiography was quite remarkable. While still maintaining religious aspects, historiography had increased its horizons and begun to treat some new subjects of general character such as, for instance, the battles between Christians and Muslims. When the ideas of the Enlightenment began to be diffused through society, it increased the velocity of transformations in historiography. What appears clearly is a new attitude from those intellectuals facing down the Catholic Church. In other words, they began to criticize church authority in their writings. At that time new scientific ideas also appeared and had a great influence on historiography by emphasizing the importance of experiments and the observation of phenomena.

Religious reform had impacted Germany significantly. Since the 16th century the teaching of a *Universal History* had been initiated in superior schools, as well as an increased necessity for educating protestant clerics and theologians. The development of this process was the birth of a *scientific history* in Germany between the end of the 17th century and the beginning of the 19th century, which coincides with the apogee of the Industrial Revolution in England with the tendency to propagate in Europe.

In the second half of the 18th century, under the influence of the Enlightenment, the necessity for a new vision of historical studies began to arise. In this context appeared Voltaire (1694–1778), whose true name was François-Marie d'Arouet, with two masterpieces: *The Age of Louis XIV* (1762) and the *Essay on the Manners and Customs of Nations* (1757). However, Voltaire is also important in the historiography of science, having written *Elements of Newton's Philosophy* (1738), which helped to popularize the theories of Newton (Voltaire 1996). With the definitive establishment of a bourgeois revolution, two prominent figures of philosophy and history emerged in Europe: Hegel (1770–1831) in Germany and Benedetto Crocce (1866–1952) in Italy. Crocce published a book about the theory and history of historiography (1917) that would become an essential a reference in this field. Both philosophers also dedicated a lot of intellectual effort to the philosophy of history.[21]

It is worth noting the influence of Positivist philosophy in general historiography as well as in the historiography of science. This school of sociological thought arose in the 19th century after the work of Saint Simon (1760–1825), August Comte (1798–1857) and Herbert Spencer (1820–1903). The entire body of work of these three authors represents more of a program for historical investigation than an investigation itself. David Émile Durkheim (1858–1917) and his followers in France and Karl Emil Max Weber (1864–1920) in Germany are considered the prominent figures of this historical thought. To explain what characterizes the positivistic proposal we can go back to the relativistic abysm to which we referred previously. It means that, in the philosophy of positivism, it is possible to separate historical facts from all ideological aspects and judgments. According to them it would be possible to elevate history to the same category as the physical sciences. Goldman criticizes positivistic thought as follows: *There was, it seems, in these researchers' thought, a not- sufficient notion of objectivity, because they attributed only to intelligence, to individual honesty of thinker, ignoring being identity with object in human sciences and their consequences to its nature and methods* (Goldmann 1970).

In spite of our description of Historical and Dialectical Materialism in 1.1, it is important to emphasize the profound impact of Marx and Engels' writings on subsequent historical thought. In addition, some historical studies in historiography must be cited owing to their importance. Marx's study of the 1848 Revolution: *The Class Struggle in France* (1850). Engels' study of the Germany movement: *Revolution and Counter-Revolution* (1852). Finally Engels' work: *The Peasant War in Germany* (1850).

It was Durkheim who formally established the academic discipline of modern social science with Marx and Max Weber. His work is mainly concerned with an epoch where traditional social and religious ties are no longer assumed. He wrote *The Division of Labour in Society* (1893) and *Rules of the Sociological Method* (1895). He was also concerned with the acceptance of sociology as a science. He then refined the positivism originally proposed by Comte and began to promote a form of epistemological realism with the use of the hypothetic-deductive model for social science.

Max Weber was a sociologist and political economist. He is perhaps best known for his thesis combining economic sociology and the sociology of religion. With respect to Weber's thoughts on their relationship to positivism, he proposed a kind of methodological antipositivism, meaning a non-empiricist field to study

[21] For an interesting study on historical and philosophical thought, see (Goldmann 1967), which articulates social and political development with dominant ideas arising in these countries in an original way.

social action. His main concern was understanding the processes of rationalization, secularization and "hopelessness" associated with capitalism and modernity. According to him modern society is a product of a new way of thinking. He wrote two famous books: *The Protestant Ethic and the Spirit of Capitalism* (1905) and *Politics as a Vocation* (1919).

Another important influence in social and historical thought in the 20th century is the Frankfurt School (Matos 1993; Slater 1978). This name is associated with the Frankfurt Institute for Social Research, which was founded in 1923 by Carl Grünberg, a Marxist and professor at the University of Vienna. Their writings clearly indicate the possibility of an alternative path to social development.

In fact, the Frankfurtians sought to find answers from other schools of thought for social transformation by using insights of antipositivist sociology, psychoanalysis, existential philosophy, and other sources. They put special emphasis on the critical component of theory that was derived from their attempt to overcome the limits of positivism, materialism and determinism, proposing then a return to Kant's critical philosophy and its successors in German idealism, mainly Hegel's philosophy based on dialectic and contradiction as inherent properties of reality.

The Frankfurt School's early members were: Max Hokheimer (1895–1973), Theodor W. Adorno (1903–1969), Herbert Marcuse (1898–1979), Friedrich Pollock (1894–1970), Erich Fromm (1900–1980), Otto Kirchheimer (1905–1965), Leo Löwenthal (1900–1993) and Franz Leopold Newmann (1900–1954) (Matos 1993).

Finally, we have to mention, because of its importance in historical studies, The School of *Annales*. In 1929, Lucien Fèbvre (1878–1956) and the great historian Marc Bloch (1886–1944) founded the magazine *Annales d'Histoire Économique* in order to continue that challenge to conventional historiography. After this first generation the name of Fernand Braudel (1902–1985) appears. The third generation arose after 1968 with Marc Ferro (1924), Jacques le Goff (1924), Andrè Berguier, Jacques Revel (1942) and finally François Furet (1927–1997) (Slater 1978).

This School combines a traditional humanistic view of history with questions and methods adopted from other disciplines and insists on a broad definition of the historian's proper field.

1.5 Science Historiography

In spite of previous recorders, the history of science as an academic and autonomous discipline appears only in the 20th century. During the classical period and in the Middle Ages it was common to associate scientific production with a historical description of previous contributions. For instance, when a Greek mathematician proposed solving a problem, he would first attack the solution through a historical evaluation of previous solutions proposed and consider this development as part of the solution (Iggers 1998).

Some ancient texts are important to the historiography of science. As example we could begin with Eudemus of Rhodes (370–300 BC), who was one of Aristotle's most important pupils. He edited his teacher's work and made it more accessible. Encouraged by Aristotle Eudemus wrote a history of Greek mathematics and astronomy. Unfortunately these texts have disappeared and only fragments of them have survived. It is only because later authors used Eudemus's writings that we still have knowledge about the early history and development of Greek science.

Another important reference is Proclus Lycaeus (412–485 AD). He was born in Constantinople and is one of the last major classical philosophers. He studied in Alexandria and wrote a commentary on the first book of Euclid's *Elements of Geometry.* This writing is one of the most valuable sources for the history of ancient mathematics. And yet, Proclus listed the first mathematicians associated with Plato.

In the middle ages, Archimedes's works were translated directly from Greek by Wiliam of Moerbeke (1215–1286), a Flemish Dominican. He was made Latin bishop of Corinth in Greece around 1286. At the request of Thomas Aquinas (1225–1274) he undertook a complete translation of the works of Aristotle.

The translations of Moerbeke were already standard classics by the 14th century. He also translated mathematical treatises by Hero of Alexandria and Archimedes. Then, he translated almost all of Archimedes works except *The Method* and *Stomachion.*

In the early 1450s Pope Nicholas V (1397–1455) commissioned Jacobus de Sancto Cassiano Cremonensis to make a new translation of Archimedes. This translation became the standard version and was printed in 1544.

Starting our modern history of science by the 18th century, Joseph Priestley (1733–1804) wrote two works with characteristics of the history of science. *The History and Present State of Electricity* (1767) and *The History and Present State of Discoveries Relating to Vision, Light and Colors (1772).* With respect to the first work, part of it is a history of electricity until 1766 and the other part is a description of contemporary theories about electricity and suggestions for future research. In the second work Priestley carefully describes the history of optics and presents excellent explanations of early optics experiments.

Priestley's science was eminently practical and he was rarely concerned with theoretical questions. He regarded the historical development as an inherent part of the development of science.[22] He lived in the period of the Industrial Revolution in England where a generalized belief in progress and the history of science was seen as part of the history of progress. This cultural situation produced biographies,

[22] Priestley is one of the most important figures of the end of the 18th century because he combines scientific investigation, humanism and politics. He discovered oxygen and, by this fact, became internationally famous. He never actually participated in politics, but because of his republican ideas, he was attacked in 1791, and his library and laboratory were destroyed. He then moved to the USA where he died in 1804 (Kragh 1989).

works about particular sciences, and writings in different fields. In this context Jean-Sylvain Bailly (1736–1793), a French astronomer wrote a history of astronomy and a series of texts between 1775 and 1782. In the same period Albrecht von Haller (1708–1777), published a collection of texts between 1771 and 1778 called *Library*. These writings were descriptions of the life and work of ancient scientists.

During the Enlightenment an optimistic but naïve concept was dominant with respect to scientific development. In addition, science was not considered a historical phenomenon. It was believed that the history of science should be studied merely as succession of chronological details, not as an integrated part of the historical field. The Scientific Revolution only expressed the necessity of knowledge for European culture, according to them. All social process was ignored.

In France, where rationalist thinking was dominant, the knowledge acquisition process was an act of pure thought, consequently abstract and a-historic. With the growth and diffusion of romantic ideas associated with strong historical character, the historiography of science was influenced by them. Then, a more relative vision of history began to develop, including the notion that the past could be judged only according to their premises.

An increasing interest in the history of science occurred due to a new form of organization and professionalization of academic life at the end of the 18th century and the beginning of the 19th century.[23],[24] At the same time instrumentalist visions about the scientific method, meaning an ideological and methodological basis to the positivism of the 19th century, began to appear and propagate. Then arose the idea of separating nature and human sciences alongside the generalized belief that philosophy must be subordinate to science. These visions and conceptions of a positivist character had penetrated established science and clearly influenced the history of science with a non-historical conception.[25]

---

[23] J. D. Bernal in (Kragh 1989) emphasizes that importance in German: *In German, specially, happened the integration of science in current academic life. German Universities had initiated their reform in the enlightenment period, in 18th century... Until eighties German Universities from different states look for founding scientific chairs and more slowly, technical laboratories whose prototype was the Liebig one, in Giessen....*

[24] The ideas of Bernal are confirmed by Eric Ashby: *In 1826, Justus von Liebig, whose eyes had been opened when he stayed in Paris to work with Gay-Lussac, created a chemical laboratory in Giessen University; it was not the first chemical laboratory in any university, but the first to offer systematic training in chemical research, and then the first nucleus of a Chemical School.*

[25] Henri Lefebvre criticizes positivism deeply, mainly in terms of how it classifies science (Bernal 1973): *The positive spirit of Comte believed to eliminate metaphysics—and he is another thing than a subtle form of metaphysics. Difficult lesson which we need to learn. It is necessary to take into account but overcoming—with dialectical relativism—this Comte´s hopeness: to eliminate the non-scientific spirit, and to constitute a worth expression, a theory of science.* Then, Lefebvre proposes another classification of science: *Among nature sciences, we need to take into account their relationships, their intermediaries: mathematical biology, biological chemistry, mathematical physics, chemical physics, etc. In the interface among human sciences and nature sciences appear human geography, the pre-history, the anthropology.*

In the eighteenth and nineteenth centuries several scientific works begin with historical introductions, becoming important historiographical sources. As an example, we can look at the historical introduction incorporated into Darwin's masterpiece *On the Origin of Species* (Lefebvre 2002) in the later editions, as we know, the first since Jean-Baptiste de Lamarck (1744–1829), a French naturalist. He proposed the idea that evolution occurred and proceeded in accordance with natural laws. Lamarck is widely remembered for the theory of inheritance of acquired characteristics.

William Whewell (1794–1866) is one of the most important figures of the historiography of science and epistemology in the 19th century. He is most known for his work in the history and philosophy of science. However, his philosophy of science was criticized by John Stuart Mill (1806–1873) such that Whewell′s philosophy appears in the 20th century only in terms of critical thought of logical positivism. Whewell is considered the first modern historian of science and epistemology with his *Philosophy of the Inductive Sciences, Founded upon their History*, published in 1840.[26] He considered himself to be a follower of Francis Bacon (1561–1626) and claimed to be renovating Bacon′s inductive method.

Further, in the 19th century, we have an important historian of science: Isaac Todhunter (1820–1884), an English mathematician. He wrote a series of histories of mathematics and physics. In fact, these texts are very specialized, difficult to be understood by anyone but specialists in the field. This is common in the majority of histories of science because they ignore methods of historical discipline. Todhunter also wrote a biographical work on William Whewell.

A more integrated perspective of science, philosophy and history is found in Ernst Mach (1838–1916), an Austrian physicist and philosopher. He agrees that a historical method is a suitable approach for gaining insight into the scientific method. Mach's masterpiece (Blanché 1983), *The Mechanics*, represents his view of the history of science. In this book he describes the function of the history of science as follows: *We shall recognize also that not only a knowledge of the ideas that have been accepted and cultivated by subsequent teachers is necessary for the historical understanding of a science, but also that the rejected and transient thoughts of the inquirers, nay even apparently erroneous notions, may be very important and very instructive. The historical investigation of the development of a science is most needful, lest the principles treasured up in it become a system of half-understood prescripts, or worse, a system of "prejudices". Historical investigation not only promotes the understanding of that which now is, but also brings new possibilities before us, by showing that which exists to be in great measure "conventional" and "accidental". From the higher point of view at which different*

---

[26] Blanché, in (1983), considers d´Alembert in his *Preliminary Discourse to Encyclopedia*, Dugald Stewart with the second volume of his *Philosophy of Human Spirit* (1814) and August Comte with his *Course of Positive Philosophy,* as epistemological precursors. He also considers *Wissenschaftslehre* (1837) by Bernardo Bolzano and *Philosophy of Inductive Sciences* (1840) by William Whewell as founding texts of epistemology.

paths of thought converge we may look about us with freer vision and discover routes before unknown (Blanché 1983).

An interesting historiography began to develop from the middle of the 19th century under the influence of several sources including Hegel, romanticism and a new historical method developed by the Berlin School: Barthold Georg Niebuhr (1776–1831) and Leopold Ranke (1795–1886). They emphasized that the past had to be analyzed on its own basis and not through contemporary visions. In addition, a systematic criticism of sources is necessary for a precise referencing. Niebuhr was a Danish-German historian also considered a father of modern scholarly historiography. Historians generally regard him as a leader of the Romantic Era and a symbol of the German national spirit. However, he was deeply rooted in the classical spirit of the Age of Enlightenment. This consideration is due to the fact that his lifetime corresponded almost exactly to the first generation of German Romanticism. For posterity, Niebuhr was the founder of the source-critical historiography (Mach 1960).

Leopold Ranke (1795–1886) was a German historian and founder of modern source-based history. He is considered the first modern historian to emphasize the importance of sources in a famous phrase: *no documents, no history.* Ranke made a huge impact on nineteenth and 20th-century historiography and many of his books became and remained standards. He wrote not only on German history but on the history of a number of states in 19th-century Europe.

Ranke's work was very popular and his ideas about historical practice gradually became dominant in western historiography. More recently his ideas of historiography and empiricism have become regarded as outdated and no longer credible. The critique of modern historians is based in the notion that historians of previous eras did not merely report facts, but actively chose which facts they would use.

Another important vision of the history of science comes from August Comte (1798–1857) in accordance with his positivistic programme which emphasized the unit of science and its relations with other parts of social and cultural life. In 1832, he proposed the establishment of a Chair of the history of science at the *Collège de France*. This Chair would be the first of its kind in the world, but it was not actually created until 1892.

In Comte's programme there are two fundamentally different ways to present or even to understand a scientific subject: the historical and the dogmatic methods. According to the latter, a scientific text is essentially ahistorical and represents a logical and clear description. From Comte's point of view, this method is necessary for philosophical and pedagogical reasons but does not contribute to the true nature of science.

Through the Comte perspective, a cultivation of progress with positivistic science, rehabilitating of medieval science, was begun contrary to the vision of the French Encyclopedists, who emphasized the specificity of the new science. However, the true contribution of Comte to the history of science was superficial. To him, the sources and historical data accumulated played a secondary role.

It was only at the turn of the century that certain activities were organized and the history of science as a discipline as well as a profession began to be established. The first international conference was held in Paris in 1900, a first step that allowed for a regular series of similar congresses to occur. Another important sign of professionalization was the appearance of national societies for the study and development of the history of science. At the same time, the first Chairs, the first regular courses and the first periodicals began to arise.

The end of the 19th century saw the emergence of Paul Tannery (1843–1904), one of the most important scientific historians (Iggers 1998). He is considered the founder of the modern history of science. Like Comte, he emphasized the history of science as being an integral part of the history of mankind. An important contemporary of Tannery's was Pierre Duhem (1861–1916), a French chemist, physician and philosopher of science. He studied the development of the physical sciences in the middle ages and the Renaissance. Because of his relation to the Catholic Church, Duhem regarded the Scientific Revolution as a mere natural development of the science of the middle ages. Although he used a rich documentation of original texts and sources, establishing a new and more rigorous history of scientific investigation utilizing arguments to highlight ths field of knowledge, his ideas of continuity were severely criticized. Two other well-known specialists with great interest in his respective disciplines were Pierre Berthelot (1827–1907) and Wilhelm Ostwald (1853–1932) (Ostwald 1887). The latter made an incredibly important contribution to physics and chemistry with a series of papers since 1889 that were collected in more than 250 volumes (Stern 1973).

Together with Duhem, the German historian of science Emil Wohwill (1835–1912) established an important school of scientific history, disciples of which are: Anneliese Maier (1905–1971), A. C. Crombie (1915–1996) and Marshall Clagett (1916–2005). They studied the predecessors and historical background of the Scientific Revolution in great detail.

At the turn of the century the renewal of the history of science was due to discoveries in archaeology, anthropology and philology. This occurred with the discovery of new historical sources extending the horizons of the history of science through revealing such unknown scientific cultures as Egyptian and Babylonian mathematics and astronomy (Iggers 1998). As example, the Danish philologist J. L. Heiberg (1854–1928) discovered in Istambul, in 1906, a manuscript that has led to a new understanding of Archimedes′s methods with respect to Greek mathematics. The Archimedian *The Method* only survived in a palimpsest discovered in 1906. The most important contribution made by the palimpsest is the method used by Archimedes to measure the volume of a cylindrical segment. At the same time as the development of new academic disciplines, the history of science became of increasing interest from an educational point of view. Many teachers begin to postulate a historically oriented method as a new pedagogical way of studying scientific disciplines. In spite of this fact, few courses practiced it.

Another name of great importance in the 20th century is the Belgian-American George Sarton (1884–1956). He was influenced by Comte (Pensadores and Comte 1991; Fédi 2000) and Tannery and tried to emphasize in his works the same view of the history of science that a belief in progress is a key element. He developed this aim in a series of articles along with attempting to organize the history of science as an academic discipline according to these guidelines (Iggers 1998). Sarton's most important contribution has been in the United States where the history of science began to be taught in some universities where the ideological climate was favorable to these visions at the end of the 19th century.

In the first decades of the 20th century Alexandre Koyré (1892–1964) arose as a kind of father and intellectual leader of a group of very famous historians of science: M. Clagett, I. B. Cohen (1914–2003), A. Crombie, H. Guerlac, A. R. Hall (1920–2009), M. Boas (1919–2009), C. Gillispie (1918), J. Murdoch (1927–2010), E. Grant (1915–1963), Thomas Kuhn (1922–1996) and Richard Westfall (1924–1996). In France Pierre Costabel (1912–1989) and René Taton (1915–2004) were also associated with Koyré. His influence was also appreciable in Great Britain.

Alexndre Koyré (Fédi 2000) was born in Taganrov, Russia into a Jewish family in 1892. He studied in Tiflis, Rostov-on-Don and Odessa before moving to Europe. In Germany (1908–1911) he studied with Edmund Husserl (1859–1938) and David Hilbert (1862–1943). In Paris, beginning in 1912, he was a disciple of Henri Bergson (1859–1941) and Leon Brunschvicg (1869–1944).

Koyré's method was based on a careful reading of historical texts as a starting point, with the subsequent absorption of the systems of creeds allowing the reader to understand and interpret a given situation. He followed the development of ideas to organize and structure those cultural patrimonies in order to make it easier for readers to enter into the intellectual world where scientists lived. He applied this method to Galileo, Kepler and Newton, the most important names associated not only with the Scientific Revolution but also with Paracelsus and Thomas Morus.

Yet, Koyré's method was developed in France of the '20 s and '30 s under the influence of both antipositivism and historicism regarding discontinuous historical approaches. His method presented two fundamental characteristics: (a) the holistic character, in which scientific thought is not independent but indeed inseparable from the global system of representation of a given time; (b) the idea that thinking systems are discontinuous, with the change of mentality being the fundamental aspect; the Scientific Revolution provides only the elements for concluding that experimentation, measurements and the mathematization of nature are a consequence of that mentality change.

From the point of view of this book, the most important works by Koyré are those related to the Scientific Revolution: *Galilaic Studies* (1940), *Nicolaus Copernicus* (1943), *A Documentary History of the Problem of Fall from Kepler to Newton* (1955), *From the Closed World to the Infinite Universe* (1957), *The Astronomical Revolution: Copernic, Kepler, Borelli* (1961), *Newtonian Studies* (1965).

Continuing with the 20th century historians of science, we find the name of Karl Popper (1902–1994), one of the greatest scientific historians and philosophers (Solis 1994). He was born in Vienna, which at that time was an important cultural centre of the western world. He obtained a primary school teaching position in 1925, took the PhD in philosophy in 1928, and qualified to teach mathematics and physics in secondary school in 1929. The dominant philosophy of science in that period was propagated by the Vienna Circle, a group of intellectuals who met under the leadership of Moritz Schilick (1882–1936). Other members were: Rudolf Carnap (1891–1970), Otto Neurath (1882–1945), Viktor Kraft (1880–1975), Hans Hahn (1879–1934) and Herbert Feigl (1902–1988).

The principal objective of the philosophers of the Vienna Circle was to unify the sciences that, in their view, implied the elimination of metaphysics by showing that its propositions were meaningless. This unification was also an attempt to reconceptualize empiricism by means of their interpretation of the recent advances in the physical and formal sciences. The anti-metaphysical purposes were supported by an empiricist criterion of meaning and a broadly logicist conception of mathematics. The Vienna Circle's life was short, from 1924 to 1936, when Schilick was murdered. In 1938, with the onset of World War II, political pressure was brought to bear against the group, and it disbanded. Many of its members fled to the United States and a few to Great Britain.

Popper then became critical of the logical positivism of the Vienna Circle. This opposition was executed systematically, resulting in Otto Neurath calling Popper the *official opposition*. The first important book published by Popper was *Logic of Scientific Discovery* (1934/35). This book presents the principal arguments broadly accepted against logical positivism.

To Popper, the central problem of the philosophy of science is that of *demarcation,* which means distinguishing between science and what he classified as *non-science*. He, unlike the majority of contemporary philosophers, accepts the validity of the Humean critique of induction, and goes beyond, postulating that induction is never actually used by scientists. However, he does not adopt the skepticism associated with Hume and also insists that pure observation as proposed by Bacon and Newton as the initial step in the formation of theories is completely misguided. To him, all observation is selective and there are no pure theory-free observations. The traditional view that science can be distinguished from non-science on the basis of its inductive methodology is not adopted by Popper. He holds that there is no unique methodology specific to science.

Popper repudiates induction and rejects the view that it is a characteristic method of scientific investigation and inference, substituting *falsifiability* in its place. A theory is scientific only if it is refutable by a conceivable event. Every genuine test of a scientific theory, then, is logically an attempt to refute or falsify it, and one genuine counter-instance can falsify the whole theory.[27]

Another remarkable place in the history of science in the first half of the 20th century is occupied by John Desmond Bernal (1901–1971), who was born in Nenagh, Ireland and died in London. He did pioneering work in x-ray crystallography, was professor of physics at Birkbeck College, University of London, and a fellow of the Royal Society. His contemporaries called him by the nickname *Sage* because of his uncommon wisdom. It is also well known that he was Marxist in philosophy and communist in politics, aligned with the thinking of the Soviet

[27] For a concise study of Karl Popper´s ideas, see (Solis 1994), which contains the relation of all Popper´s books published in France.

Union. In this context, he played a leading role among scientific and political organizations.

Bernal was first educated in England. In 1919, he went to Emmanuel College from Cambridge University with a scholarship. At Cambridge Bernal studied both mathematics and science for a B.A. degree in 1922. In 1927, he was appointed as the first lecturer in Structural Crystallography at Cambridge, becoming assistant director of the Cavendish Laboratory in 1934.

Bernal's most important works were *The Social Function of Science* (1939), probably the earliest text on the sociology of science, and *Science in History* (1954), a four-volume attempt to analyze the interaction between science and society. We should mention *The Origin of Life* (1967), that worked with the contemporary ideas of Aleksandr Oparin (1894–1980) and John Burdon Haldane (1892–1964).

Bernal became a prominent intellectual in the political life of Great Britain in the '30 s. He attended the Second International Congress of the History of Science in London in 1931, where he met the Soviet delegation under the leadership of the revolutionary Nikolai Bukharin (1888–1938). In this congress Boris Hessen (1893–1936) presented a paper that became famous in the history of science: *The Socio-Economic Roots of Newton's Principia.* Hessen's thesis was that intellectual achievements such as Newton′s theory of motion are best explained by examining the social context out of which they arose.

Passing to the second half of the 20th century, we find a very influential historian of science, Thomas Samuel Kuhn (1922–1996). He began his academic career as a physician but certain circumstances led him to the history and philosophy of science. He was born in Cincinnati, Ohio, United States. Kuhn obtained his B. S. degree in physics from Harvard University in 1943, and MSc and PhD, also in physics, in 1946 and 1948, respectively. In this period he switched from physics to the history and philosophy of science. After Harvard he moved to Berkeley, University of California, and was named Professor of the History of Science in 1961.

At Berkeley, Kuhn wrote his most influential work (Bouveresse 2000): *The Structure of Scientific Revolutions* (1962) (Kuhn 1945). He successively joined Princeton University and the Massachusetts Institute of Technology. His masterpiece was originally a paper featured in the International Encyclopedia of Unified Science, a publication of logical positivists from the Vienna Circle. In Kuhn's view, science does not progress as a linear accumulation of new knowledge. To him, science is transformed by periodic revolutions, also called *paradigms shifts*, a phrase not coined by him, in which a particular field of knowledge is abruptly changed. In general, this rupture has three distinct stages.

The first is a prescience, which lacks a central paradigm. This stage is followed by *normal science* when the scientific community attempts to enlarge the central paradigm. This period is the most productive in terms of the solution of problems. When some anomalous result appears, it is attributed to a mistake of the researcher. During a normal scientific stage, scientists neither test nor seek to confirm the guiding theories of their disciplinary matrix.

It is only the accumulation of particular anomalies that poses a serious problem for the existing paradigm. If this problem represents a widespread failure in confidence with respect to the paradigm, we have a *crisis*. Naturally the response to a crisis will be the search for a revision of the guiding theory. Here, there is an interesting intellectual dispute between Kuhn and Popper. According to Popper, the revolutionary overthrow of a theory is one that is logically required by an anomaly. According to Kuhn, there are no rules for deciding the significance of one solution against another. For this reason, the revolutionary phase is particularly open to competition among different ideas. Some sociologists and historians of science enter into that discussion and suggest that not only the revolutionary stage but any step in the development of science is always determined by socio-political factors. Kuhn repudiated these ideas and believed that the solution to this problem can be found within science.

We have to mention that Kuhn wrote other important works for the purpose of our book: *The Copernican Revolution: Planetary Astronomy in the Development of Western Thought* (1957); *The Essential Tension: Selected Studies in Scientific Tradition and Change* (1977), which contains an interesting history of the principle of conservation of energy, established in 1847 (Kuhn 1957; Kuhn 1996).

Allen George Debus (1926–2009), known primarily for his contribution to the history of chemistry and alchemy, was born in Chicago, United States. He studied chemical engineering and history, graduating in chemistry in 1947 from Northwestern University. He obtained his MSc degree at Indiana University in 1949, presenting a thesis on *Robert Boyle and Chemistry in England* 1660–1700. In 1956, he finished his PhD thesis at Harvard, guided by the famous historian of science Bernard Cohen. In 1959, he went to London, and attended University College of London courses given by Douglas Mckie. Then, he returned to Harvard where he completed the requirements for his PhD in the history of science.

The book *Man and Nature in the Renaissance* (1978) is Debus′s most important contribution to understanding science in the Renaissance. Frequently, the approaches used only describe the progress of the exact sciences of mathematics and astronomy, ignoring the broader intellectual context of the period. Thus, he renewed interest in mystical texts and the subsequent impact of alchemy, astrology and natural magic on the development of modern science. Other important books published by Debus are: *The English Paracelsians* (1965); *The Chemical Dream of Renaissance* (1968); *Chemical, Alchemy and the New Philosophy,* 1550–1770: *Studies in the History of Science and Medicine* (1987); *Chemistry and Medical Debate: van Helmont to Boerhaave* (2001); *The Chemical Promise: Experiment and Mysticism in the Chemical Philosophy,* 1550–1800 (2006).

The final historian of science presented in this chapter is Helge Kragh, professor of History of Science and Technology, Department of Science Studies, Aarhus University, Aarhus, Denmark. He was president (2008–2010) of the European Society of the History of Science. Kragh's academic interests are in the history of modern physical sciences, including chemistry, astronomy, physics and cosmology, as well as the philosophical aspects of science and science-technology relations.

We come back to the debate of ideological aspects in history, particularly in the history of science. Kragh's book *An Introduction to the Historiography of Science* (1989) introduces the methodological and philosophical problems with which the modern history of science is concerned. With respect to ideological problems, Kragh wrote: *Histories of science involve particular perspectives, aims and methods of organizing materials that do not arise out of the objectively given past itself. Very often, history of science also serves legitimating functions. The fact that histories are written with commitment and from a particular motive, or may serve legitimating functions, does not necessarily imply that they are products of bad historiography. But as soon as documentary evidence is distorted, ignored or allocated disproportionate importance in order to fit in better with a particular moral that serves a social function, history becomes ideological (Iggers* 1998*).*

Kragh also gave some examples of the above situation: *During and immediately after the First World War, the hostility between the belligerent parties resulted in histories of science that were markedly nationalistic. For example, the eminent French physicist and mathematician Émile Picard wrote a history of science in 1916 that was to show that all that was good in the development of science was due to French scientists and all that was bad to German Scientists.* Another example is the history of science in the Soviet Union: *Nationalistic mythicization received a special stamp in the Soviet Union from about 1930–1955. History of science was used ideologically, as defense of the political system and in order to increase Russian national pride. It was hoped to help counteract the Soviet feeling of cultural and scientific backwardness by the use of a history of science designed for the purpose. This history was marked by, among other things, xenophobia and the assertion of a series of priority claims* (Iggers 1998).

References

Althusser L (1967) Análise Critica da Teoria Marxista. Zahar Editores, Rio de Janeiro

Anderson P (1976) Considerações sobre o Marxismo Ocidental. Edições Afrontamento, Porto

Bernal JD (1973) História Social de la Ciência. Ediciones Península, Barcelona

Blanché R (1983) A epistemologia. Editorial Presença, Lisboa

Bodanis D (2000) $E = mc^2$: Uma Biografia da Equação que Mudou o Mundo e o que ela Significa. Ediouro Publicações, Rio de Janeiro e S. Paulo

Bouveresse R (2000) Le Rationalisme Critique de Karl Popper. Ellipses Édition Marketing, S. A., Paris

Burkett P (2009) Marxism and ecological economics. Haymarket Books, Chicago

Cohen GA (2000) Karl Marx's theory of history. A defence. Clarendon Press, Oxford

da Silva RF (2000) História da Historiografia. Editora da Universidade do Sagrado Coração, Bauru

Darwin C (1859) A Origem das Espécies e a Seleção Natural, 5ª Edição. Hemus Editora Limitada, São Paulo

Engels F (1962) Do Socialismo Utópico ao Socialismo Científico e Ludwig Feurbach e o Fim da Filosofia Clássica Alemã. Editora Fulgor, S. Paulo

Engels F (1966) Dialectics of nature. Progress Publishers, Moscow

Engels F (1971) Anti-Dühring. Edições Afrodite, Lisboa

Fataliev K (1966) O Materialismo Dialético e as Ciências da Natureza. Zahar Editores, Rio de Janeiro
Fédi L (2000) Comte, Le Belles Lettres. Figures du Savoir, Paris
Goldmann L (1967) Origem da Dialética: a Comunidade Humana e o Universo em Kant. Editora Paz e Terra, Rio de Janeiro
Goldmann L (1970) Ciências Humanas e Filosofia. Difusão Européia do Livro, São Paulo
Herôdotos (1988) História. Editora da Universidade de Brasília, Brasília
Hobsbaum EJ (1980) História do Marxismo, 12 volumes. Editora Paz e Terra, Rio de Janeiro
Iggers GG (1998) La Ciência Histórica em el Siglo XX. Idea Books, S. A., Barcelona
Kragh H (1989) An introduction to the historiography of science. Cambridge University Press, Cambridge
Kuhn TS (1957) A Revolução Copernicana. Edições 70, Lisboa
Kuhn TS (1945) A Estrutura das Revoluções Científicas. Editora Perspectiva, São Paulo
Kuhn TS (1996) La Tension Esencial. Editora Fondo de Cultura Economica, México
Labica G (1990) As Teses sobre Feuerbach de Karl Marx. Jorge Zahar Editor, Rio de Janeiro
Lefebvre H (1940) Le Materialisme Dialectique. Quadrige/Presses Universitaires de France, Paris
Lefebvre H (2002) Methodologie des Sciences. Anthropos, Paris
Lênin VI (1971) Materialismo e Empirocriticismo. Editorial Estampa, Lisboa
Löwy M (1987) As Aventuras de Karl Marx contra o Barão de Munchausen. Editora Busca Vida, S. Paulo
Löwy M (1988) Ideologias e Ciência Social: Elementos para uma Análise Marxista. Cortez Editora, São Paulo
Lukács G (1971) A Falsa e a Verdadeira Ontologia de Hegel. Livraria Editora Ciências Humanas LTDA, São Paulo
Lukács G (1979) Existencialismo ou Marxismo?. Livraria Editora Ciências Humanas LTDA, São Paulo
Mach E (1960) The science of mechanics: a critical and historical account of its development, 9th edn. The Open Court Publishing Company, Illinois
Markus G (1974) A Teoria do Conhecimento no Jovem Marx. Editora Paz e Terra, Rio de Janeiro
Marx K (1971) Contribuição para a Crítica da Economia Política. Editorial Estampa, Lisboa
Marx K (1972) Diferença entre as Filosofias da Natureza em Demócrito e Epicuro. Editorial Presença, Lisboa
Marx K, e Engels F (1980) A Ideologia Alemã. Editorial Presença, Lisboa
Marx K, e Engels F (1998) Manifesto Comunista. Boitempo Editorial, S. Paulo
Marx K, Engels F (1965) Selected correspondence. Progress Publishers, Moscow
Matos CFO (1993) A Escola de Frankfurt. Editora Moderna, Coleção Logos, Coleção Logos
Ostwald FW (1887) Journal of physical chemistry. Leipzig, Germany
Paraire P (1997) Autour d'Épicure. Le Temps de Cérises, Paris
Pensadores Os (1991) AugustoComte. Vida e Obra. Editora Nova Cultural, São Paulo
Piaget J (1997) Biologia y Conocimiento. Siglo Veintiuno Editores, Madrid
Políbios (1996) História. Editora Universidade de Brasília, Brasília
Prigogine I (1996) Tempo, Caos e as Leis da Natureza. Editora Unesp, S. Paulo
Slater P (1978) Origem e Significado da Escola de Frankfurt. Biblioteca de Ciências Sociais, Zahar Editores, Rio de Janeiro
Solis C (1994) Alexandre Koyré. Ediciones Paidós, Instituto de Ciências de La Educacion, Barcelona
Stachel J (2001) O Ano Miraculoso de Einstein. Editora UFRJ, Rio de Janeiro
Stern F (1973) The varieties of history. Vintage Books, New York
Tétard P (2000) Pequena História dos Historiadores. Editora da Universidade do Sagrado Coração, Bauru, São Paulo
Tucídides (1999) da Guerra do Peloponeso. Editora Universidade de Brasília, Brasília
Voltaire (1996) Elementos da Filosofia de Newton, Editora da Unicamp, Campinas, São Paulo

Part I
The Conceptual Genesis

Chapter 2
The Conceptual Basis to Work Studies

"**Socrates**—Can it be, then, that Protagoras was a very ingenious person who threw out this dark saying for the benefit of the common herd like ourselves, and reserved the truth as a secret doctrine to be revealed to his disciples?

Theaetetus—What do you mean by that, Socrates?

Socrates—I will tell you; and indeed the doctrine is a remarkable one. It declares that nothing is *one* thing just by itself, nor can you rightly call it by some definite name, nor even say it is of any definite sort. On the contrary, if you call it *large*, it will be found to be small; if *heavy,* to be also light; and so on all through, because nothing is *one* thing or *some* thing or of any definite sort. All the things we are pleased to say *are,* really are in process of becoming, as a result of movement and change and of blending one with another…"

(Plato's Theory of Knowledge—The Theaetetus and the Sophist—Translated by Francis M. Cornford—p. 36, Dover Publications, New York, 2003)

2.1 Historical Character of Physical Concepts Construction

In his paper: The historical character of adequacy of mathematics to physics, Michel Paty[1] wrote: "Following through history of sciences, several remarkable cases of relations between physics and mathematics, we realized that the capacity of mathematics formalism to express in an appropriate and fruitful manner the physical problems, is not a fact of universal and extemporal nature. It is the result, in each time, and to each new studied problem, a construction, that put in action the "system" of the mathematics and physics of this time and the nature of the concepts and physical quantities involved."

By analyzing this chapter with respect to a conceptual framework in which the concept of work is supported, we made an initial decomposition of the elements

An erratum to this chapter is available at 10.1007/978-94-007-7705-7_9

[1] Michel Paty specifically treats the problem of mathematics' adequacy to physics in two papers (Paty 1991, 1999). See also Petitot in (Petitot 1990).

A. R. E. Oliveira, *A History of the Work Concept*, History of Mechanism and Machine Science 24, DOI: 10.1007/978-94-007-7705-7_2,

of work, complemented by a historical perspective in a similar way made by Paty as cited above, meaning that the concepts of physics and the utilization of mathematics in so fruitful form belong to a common mathematical-physical construction of historical character. In addition, one also may conclude that it is necessary for each particular physics' problem to mobilize all available mathematical tools, or, if necessary, to develop it. However, one recognizes that there is a fundamental constitutive aspect of mathematics with respect to the representation of the physical world that is a more intrinsic indication of the participation of mathematics in it.

Obviously, this conclusion is not so explicit from the citation, but one can conclude this from a reading of the text. It means that mathematics is not a mere language used for representing reality or to operate symbolically, as some authors postulate. In other words, mathematics belongs to a concept's construction, although the citation also emphasizes that there is a kind of symbiosis between mathematics and physics and that it is not a general or universal fact independent of the physical problem under consideration. This would be equivalent to considering mathematics as a tool ever available to physics, which is not true. What, in fact, happens is a complex relationship of a common construction in both fields but historically conditioned.

The utilization and adequacy of mathematics as a better form of representation of the physical world are appropriate in well-defined moments, historically dated and also referring to precise theories and forms as well as to defined types of physical problems.[2] Particular to the field of Rational mechanics, for example, it is well known that mechanical laws have profited significantly from the invention of differential and integral calculus developed some years before by Newton and in an independent form by Leibniz.

The formulation of fluxions calculus by Newton, as a result of his achievements in infinite series, is simultaneously the study of the motion problem of bodies considered according to geometrical methods which also led to a new branch of mathematics (Newton 1994). Newton was aware of how physics influenced the construction of mathematical concepts when he stated in the preface of the first edition of *Principia*: *Geometry is founded on mechanical practice, and is this part of universal mechanics that postulates and demonstrates exactly the art of measurement* (Newton 2002).

In order to corroborate the above statement, Michel Paty also considers that the foundation of calculus invented by Newton in 1666 uses mathematical quantities that are generated by a continuous motion which opposes concepts of infinitesimal quantities. Indeed, Newton does not directly use differential calculus in his mechanics, especially in the derivation of his second law of motion, which would be done some years after by Leonhard Euler. The Swiss mathematician formulated the mathematical expression of Newton's second law in 1752.[3]

---

[2] Michel Paty in (Paty 1991) presents some examples of mathematical constructions of physical concepts in two fields: classical mechanics and the theory of relativity.

[3] Euler, in *The Discovery of a New Principle of Mechanics*, published in 1752, presents the differential equations expressing Newton´s second law. Later on, he called these equations *the first principles of mechanics* (Truesdell 1983).

Again, Paty emphasizes that Newton remarks in his *first and last reasons*, which open book I, *The bodies' motion*, are quite similar to what is found in the calculus of fluxions. He wrote: "The conceptualization of fluxions—without the symbolism—penetrates in some way this geometry transforming it in relation to mechanics, what will be translated in an evident manner by the use of differential symbolism (Paty 1991)". In other words, the absence (explicit) of calculus in Newtonian mechanics is not a complete truth and represents a more complex process which Michel Paty realizes and remarks upon with great perspicacity.

The formalization of mechanics which begins with Pierre Varignon and d'Alembert, as we will see later on, continues with Euler and has its crowning with Lagrange; it will be presented in detail in Chap. 4. Besides these important names, Jacques and Jean Bernoulli developed and applied this new methodology to mechanics.

In fact, this period is very rich in the construction of basic mechanical concepts and illustrates practically what was discussed above concerning the relation between mathematics and representation of the physical world. However, only with the generation of mathematicians of d'Alembert and Euler, but incorporating Daniel Bernoulli, Alexis Clairaut and others, can we say that the universalization of differential and integral calculus has happened. This fruitful process opens multiple possibilities for the development of mechanics and astronomy, mainly, exploring Newton's system of the world beyond the limits of the studies of bodies' motion. The way adopted by Lagrange via mathematical analysis led him to the great mechanics' formalization in 1788 with his *Analytical Mechanics*.

Michel Blay, in his *La Naissance de la Mécanique Analytique* (Blay 1992), made an analysis similar to our previous description concerning mechanical development with the creation of differential and integral calculus, also with respect to the role of mathematics in the description of the physical world. Thus, he states: "The comparison between leibnizians and varignonnians methods in the second part, has shown us clearly that the differential conceptualization of the science of motion cannot be interpreted as result of a mere transposition in differential terms, of the concepts and view points of the science of motion such that it was expressed previously, but as the result of a true conceptual reconstruction that has led to a reorganization of the science of motion field around the concepts of instantaneous velocity and acceleratrix force."

He remarks the following about the relationship between mathematics and physics: "From a general point of view, the conceptual varignonnian work testifies that mathematics and physics have a constitutive relationship, or better, a constituent relationship, because by analyzing this work, in the dynamics of this relation—one can say also about its interaction—that they are intrinsically related to the new concepts, in the occurrence here, of the instantaneous velocity and acceleratrix force".

Another important aspect highlighting the historical character of the formation of concepts to the construction of theories, mainly those concerning the physical world, is technological development. As we know, concepts and theories must demonstrate their explanatory capacity as well as their fruitfulness facing the facts (phenomena) with respect to the physical world. Therefore, they need experimental validation, which only happens after a long process of improvement of the tools and experimental techniques used. As an example, we can use the restricted theory of relativity that arose after experimental demonstration of the constancy of light velocity related to any system of reference. This fact implied a reconstruction of mechanical theory with serious consequences for ether theory that became cumbersome in spite of resistance.

An example of how an experimental apparatus interferes with theories representing reality comes from models of matter constitution,[4] which need sophisticated particle accelerators, as well as very specialized computational technologies. From the informatics point of view, the use of computers in physical studies has introduced a conceptual revolution. It was possible also to discover unknown aspects of the behavior of systems, such as chaotic behavior in deterministic systems (Prigogine 1996) transforming classical mechanics into a modern tool for system analysis. As a consequence of these modifications, several new concepts have arisen: fractals, strange attractors, complexity phenomena (Nussenzveig 1999) and a new branch of geometry with non-integer dimensions. However, besides the new concepts, some new scientific disciplines have also appeared: graphics computation, artificial intelligence, etc. On the other hand, these new tools create favorable conditions for new discovery. Finally, it is worth noting, as we said in the first chapter about the theory of knowledge, that men historically construct their tools to explore nature and that our approach to reality is progressive, neither continuously nor linearly, but with the rhythm dictated by the capacity to construct these tools.

By analyzing the evolution of physical concepts, one observes that they are historically conditioned to technological development, so obviously it is not the unique factor. This evolution also depends upon the development of other concepts and theories that in some cases have a crucial importance to the progress of physics. One can remark that mechanical theory has known a fruitful development after the application of differential and integral calculus. Calculus' formalism has made the manipulation of algebraic problems, primarily those after Euler, much easier. The success of this mathematization has induced a very fruitful mathematization in other particular sciences, such as heat, electricity, magnetism, optics, etc., in spite of the failure of the Laplacian project to extend the Newtonian model of gravitational theory to the other sciences.

The basic concepts or elementary concepts for studying the concept of work had a historical development. This is the objective of the following sections.

---

[4] The experiments with modern particle accelerators can explore scales up to 10^{-18} m. A new accelerator is now in operation, the LHC (Large Hadron Collider). It was built by CERN (European Organization for Nuclear Research) from 1998 to 2008, and is the largest and highest-energy particle accelerator to date, with the objective of testing the predictions of different theories of particle physics and high-energy physics, particularly for the existence of the hypothesized Higgs boson and the large family of new particles predicted by supersymmetry (Special edition of Scientific American 2005; D'Espagnat and Klein 1993).

2.2 The Evolution of Space Concepts

In this topic, we limit our analysis to the presentation of the main conceptions of space directly related to the concept of work. They come from Euclidian geometry as well as the visions of the founders of mechanics concerning space.

2.2.1 Euclid (330–260 B.C.)

Euclid's *Elements* were published around 300 B.C. and undoubtedly had fundamental importance in numerous fields of knowledge, becoming a paradigm of knowledge in general. To emphasize its importance we can add that Newtonian mechanics and the special theory of relativity use Euclidian space as an essential part of their theories.

Before the appearance of *Elements*, geometry was not a completely systematized science. Geometrical knowledge of the ancient peoples were dispersed and disorganized. Egyptians knew several geometrical properties and how to apply them to many problems but they ignored logical relationships that linked and unified this knowledge. In other words, before Euclid the validation of geometrical knowledge occurred only empirically dictated by practical necessities. Egyptian knowledge was adopted by the Greek people, who developed it significantly. Thales, as an example, brought to Greece what he had learned from Egyptians in the seventh century B. C. This interchange of knowledge among several cultures with the development added by the Greeks permitted the establishment of logical connections within geometry. In addition, it became clear that geometrical knowledge is based on basic principles. Euclid realized this tendency and thus he collected and organized proofs, classifying the material and adding his contributions. Adopting this approach he concluded and demonstrated in his *Elements* that all geometrical knowledge is obtained through deductive method starting from five postulates and its consequent deductions. With this operation, the whole of geometry will be confined into the axioms from which the theorems are derived .[5]

These preliminary remarks lead us to say that the importance of Euclid is not only through the discovery of what postulates are true but even more so through the revelation of the five axioms that lead to geometry as a whole. Later on, David Hilbert, at the end of the nineteenth century, showed that Euclidian geometry can be derived through 20 axioms. He added one more, the completeness one. It is

[5] Leonard Mlodinov (Mlodinow 2004), p. 49, explains Euclid´s contribution: The most important Euclide´s contribution in "The Elements" was his logical and innovative method: first, by explicitating the terms, by formulating precise definitions and guaranteeing thus the understanding of all the words and symbols. In addition to become explicit the concepts presenting clearly the axioms or postulates (these terms are interchangeable) such that it cannot be used with understanding or suppositions not declared. Finally, to derive logical consequences from the system using only logical and accepted rules, applied to axioms and theorems previously proved.

Fig. 2.1 Latin translation of *the Elements* (twelveth century)

remarkable to note that in modern formulations the original five Euclidian postulates have accomplished a great number of derivations, and, because of their simplicity, they continue to be used nowadays .[6]

In terms of space theory, Euclid's *Elements* present the basic principle of his theory in the form of definitions and postulates, which determine an abstract world that, in the case of two dimensions, represents a plane. From this starting point, he derives the theorems in order to discover other evidence giving consistency to this geometric world. This means that deductions are logically valuables, and, if the axioms are true, the theorems are also true (Fig. 2.1).

It is not our objective to submit Euclidian geometry to epistemological analysis in order to study its consistency as scientific theory. However, it is important to discuss a fundamental concern of all theory that is its adequacy as a representation

[6] Concerning the logical consistency of Euclidian geometry or any axiomatic system, the history in general is as follows: In 1903, the philosopher and mathematician Bertrand Russell in his book, "Mathematical Principles", suggested that the whole of mathematics would be derived from logic. He tried with Alfred Whitehead to show how to do this in the book "Principia Mathematica", in three volumes, published between 1910 and 1913. In this book, the authors have affirmed the reduction of all mathematics to a unified system of basic axioms from which all theorems of mathematics could be demonstrated in a similar manner as Euclid tried to do with geometry. However, in 1931, Kurt Gödel demonstrated that statements that cannot be demonstrated must exist in an axiomatic system. As a corollary of this famous theorem, at least one true proposition that cannot be demonstrated must also exist in the axiomatic systems (El Nidditch 1995; Nagel et al. 1989).

of the real world. First of all, the axioms of Euclidian geometry being a set of propositions, this fact, in principle, would induce us to think that this theory could be false. It is not the case. A great number of mathematicians and philosophers have carefully analyzed this question and concluded that Euclidian geometry is a correct theory to describe physical space. This also implies another complementary question, also fundamental, that is, what experimental evidence should this theory present in addition to logical internal coherence? Thus, the next step will be to build experiments in order to test the predictions of the theory.

Obviously, if we should submit Euclidian theory to an experimental test, it would be by means of theorems. As an example, we examine the theorem of *internal angles* of a triangle. We can build a triangle in space, measure the internal angles and verify if its addition is equal to two straight angles according to the theorem.

At the beginning of the nineteenth century, the mathematician Karl Friedrich Gauss (1777–1855) accomplished a kind of experiment very similar to that which we are imagining. A question could be asked. Why was this type of test never done? Probably because, until that time, nobody had seriously considered the possibility of our physical space being non-Euclidian. In the experiment assembled by Gauss, the natural assumption that light beams travel in a straight line within the space was made. Thus, he has measured the internal angles of a spacial triangle the sides of which were light beams. This was done by assembling the equipment on the tops of three mountains, simulating a triangles' vertices. After measurements, Gauss confirmed the Euclidian assumption .[7]

2.2.2 Descartes (1596–1650)

The fundamental basis of Cartesian philosophy is the search for true knowledge, beginning with the decision *to be doubtful about all things found with minimum possibility of uncertainty.* This statement can be read in: *The Principle of Philosophy* (Descartes 1997) (Fig. 2.2).

From the historical point of view, Descartes is the father of modern philosophy and one of those responsible for the fatal blow against medieval philosophy, theology and dogmatism. His major accomplishment, as described by Roland Corbisier: "What did Descartes, his founder? (Of the modern philosophy) He does not start from the certainty, but on the contrary, from the doubt. He does not accept any dogma, any truth, unless it can be submitted to a critique, to a reason critique, becoming free from religious domination and theological alienation, rebuilding the autonomy and the right to be the last instance of truth. The rebirth of philosophy implied with the rupture of dogmatism, or, in other words, this rebirth was conditioned by the recuperation of freedom, by the reconquering of the right to think, to

---

[7] For a complete study of Euclid´s work, see (Heath 1956).

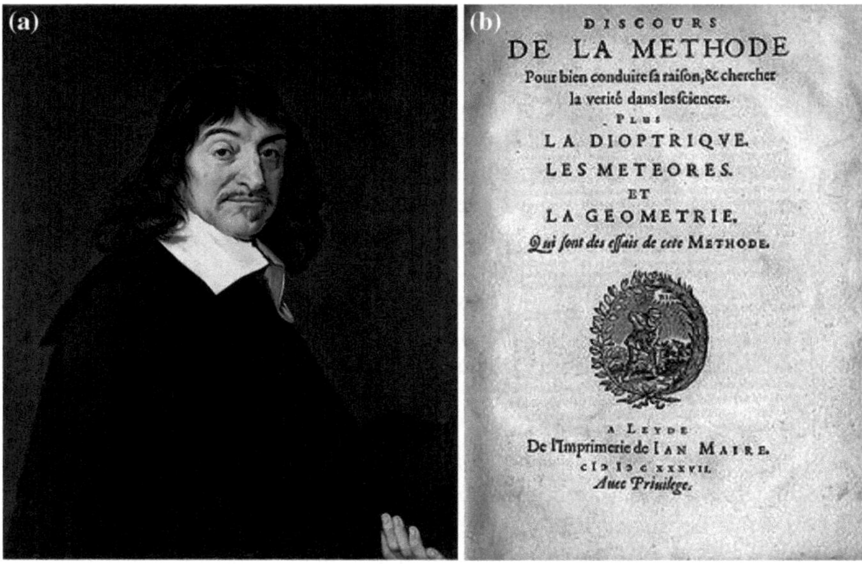

Fig. 2.2 **a** René Descartes (1596–1650), **b** First edition of Discours *de la Méthode*

criticize, to disagree. The first initiative, in Cartesian procedure, consists in eliminate of his belief all received opinion in order to replace that beliefs by others, betters, or the same opinions, since that be modified or "adjusted according to the reason".

The central thesis of Cartesian methodology is that knowledge is achieved through an act of the mind and not by experience: we can take into account many ideas and we can use our mental power to determine what is true. Such methodology, which is emphasized by Plato and Leibniz, is well known by rationalism. The opposite thesis, as we know, is empiricism, which is found in Aristotle and especially in Berkeley and Mach, where experience provides the basis for true knowledge.

Another important characteristic of Descartes' thought, for the study of the conception of space, is the distinction between mind and matter or material substance, that is, the difference between the thing that thinks and the thing that fulfills the space. To Descartes, all things except God are made by these two substances or understood by means of them. This concept of Descartes' was important at that time because things related to matter could be the object of scientific studies but only religion was permitted to be dedicated to matters of the mind, or the soul, without any interference in the progress of scientific development, and vice versa.

Descartes believed that the essential characteristic of matter is extension in three directions. He rejected such properties as hardness, weight, color and other properties that could define the characteristics of matter. This concept is fundamental to his understanding of space. To him, space is pure extension and, in addition, space and matter are the same thing. Obviously, to Descartes the idea of

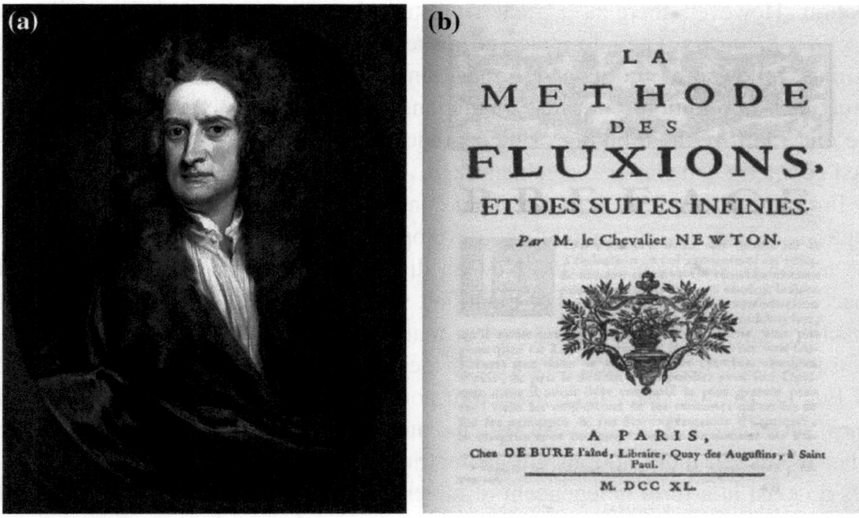

Fig. 2.3 **a** Isaac Newton (1642–1727), **b** French translation of *fluxions methods*

empty space is complete nonsense because empty space cannot exist. And yet, to him, Euclidian geometry correctly describes space if conceived according to his conception. However, the fundamental contribution of Descartes to scientific thought is obviously *Geometry*. This work, which is composed of other studies called Essays, appeared in 1737, when Descartes was 40 years old. Initially, this publication was known as *Discourse on Method*, with the sections on dioptrics, and the studies on meteors and geometry, added as appendices[8].

2.2.3 Newton (1642–1727)

It is worthwhile, in order to have a better understanding of Newton's work, to establish a comparison between both mathematicians, or, in other words, to compare Newton's *Principia* and Euclid's *Elements*. As we know, Newton's *Principia* is the fundamental reference of classical mechanics, and was published in 1687, while Euclid's book represents the original source of classical geometry, and was published around 300 B. C. (Fig. 2.3).

Euclid systematized geometrical knowledge, and Newton, in addition to this role with respect to mechanical knowledge, discovered the fundamental laws of

[8] Newton defines absolute space as follows: *The absolute space, in his nature, without relation with any external thing, stays always similar and stopped. Relative space is some dimension or moving measurement of absolute spaces to which our senses determine by his position with relation to bodies, and is usually considered not moving space...* (Newton 2002), p. 45

motion. However, there is a great similitude between Newton's *Principia* and Euclid's *Elements*. Their formats and presentation in the form of definitions and axioms, followed by proofs and propositions, are quite similar, as are their logical structures. In addition, both mathematicians were true investigators and followed the clues left by the thinking of their predecessors. Thus, the genius of both created geometry and classical mechanics.

From the point of view of our main concern, that is, the study of spacial conceptions in order to understand the development of this concept associated with the notion of mechanical work, Newton disagrees with Descartes' conception of space. To Newton, matter is not identified with space; his alternative vision is of absolute space and its independence of matter. Hence, Newton attributed a status of true knowledge to space as a kind of geometrical object, exactly as conceived by Euclid. To him, space is composed of points, a collection of points constructing lines, surfaces and solids having volume, and not properties of material bodies. In other words, all these notions are geometrical properties. These geometrical entities do exist in a form independent of material objects and their locations do exist before being occupied by material objects.

Space, to Newton, is also absolute in the sense that it exists even if there is no object inside of it. Its existence is independent of matter and it was in this sense that Newton thought of absolute space[9].

Unlike Descartes, Newton believed that space is quite different from matter in a similar way that objects are different among themselves. Thus, points, lines, surfaces and solids do exist and persist through time remaining the same size. On the other hand, absolute space is infinite, tridimensional, and rigid, as if a kind of Euclidian box that exists, immutable with time.

Newton also elaborated the concept of relative space which considers the position of an object in relation to another adopted as a reference. As an example, to fly a plane we need to know our relative position to a road. In other words, it is necessary to define a system of reference relative to a given object. In this reference system, Newton introduced the concept of relative motion[10].

---

[9] Newton distinguishes absolute and relative motions: *Absolute motion is the translation of a body from an absolute place to other; relative motion, the translation from one relative place to other. Thus, in a ship in motion, the relative place of a body... is that part of ship that is occupied by the body. But real not moving, absolute, is the permanence of body in the same part of that not moving space, in which the ship, its cavity and all that it contains, is moving... Because the parts of space cannot be seen or differentiated one with relation to others by our senses, we use realized measures by them. Although, from the positions and distances to things from any body considered not moving, we define all places, and then, with respect to such places, we estimate all motions, considering bodies as transferred from some places to others. Thus, instead of places and absolute motions, we use the relatives, without any practical inconvenience...* (Newton 2002), p. 45.

[10] This argument is emphasized by the publication of B. Russel´s book in 1900 about Leibniz philosophy. The author postulates that Leibniz philosophy was practically derived from his logics (Parkinson 1995).

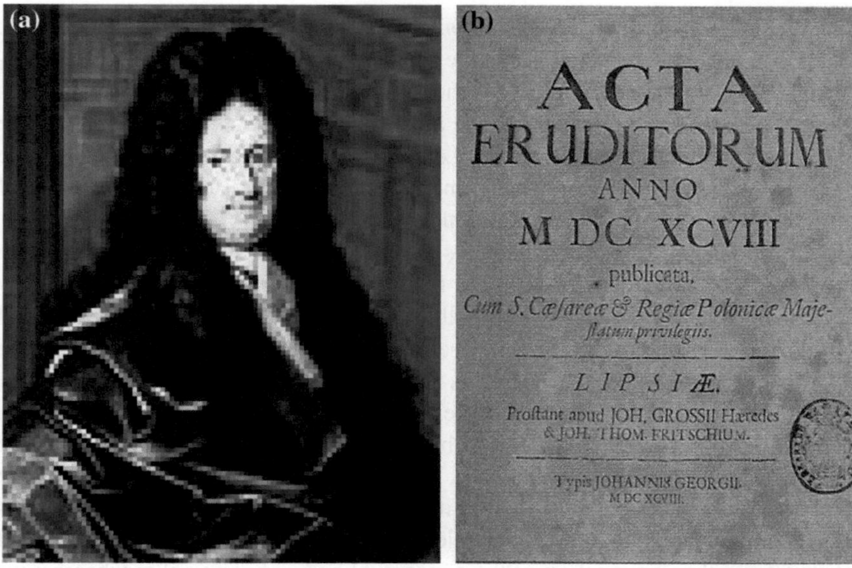

Fig. 2.4 **a** Leibniz (1646–1716), **b** Issue of Acta Eruditorum

2.2.4 Leibniz (1646–1716)

Like Descartes, Leibniz belongs to the rationalist school of philosophy. To him, geometry was a science related only to reason, and not dependent on any kind of experimentation for validation. Rationalists in general have applied this model to all knowledge and also defended their point of view based on basic axioms that permit the derivation of all facts in the real world.

Leibniz tried to accomplish an intellectual task of greater amplitude than Euclid. He postulated the idea that it may be possible to derive any knowledge (non-mathematical) from a general axiom, denominated in *Principle of sufficient reason* (Fig. 2.4)[11].[12]

If we look at Leibnizian conceptions of space, it is possible to characterize them as *relational*, Leibniz being the first thinker to present basic arguments and objections to the conceptions of absolute space and time proposed by Newton. According to Leibniz's vision, space is not an absolute substance but a *relational* category. This

---

[11] According to Ross: *Leibniz´s approach tries to reconcile logical traditions, rhetoric and mathematics by putting together three distinct things (in formalism, in the linguistic property and in the mathematization) in a unique vision of a formal language of mathematical notation* (Ross 2001) .

[12] According to Daniel Garber, p. 302 of (Parkinson 1995), of the paper *Leibniz physics and philosophy, one reads: Leibniz was concerned with the notion of space in his early writings. Hence, there is a strong suggestion in some of these writings that he thought about space as something distinct from the body. In his writings from maturity he clearly denies the reality of space independent, particularly in opposition to the specific form in which it was presented by Newton´s writings.*

means that there is an *order* or a *situation* where bodies are related to themselves[13]. Looking at a relative reference system, we can determine relative positions of all bodies by taking one of them as reference, although any object can, in principle, be used as reference, such that it is possible to have a collection of objects with several relative distances to each other. As an example, Leibniz said that we can have the sun or the earth as systems of reference to the planets, being that there are different ways of describing relative distances among all of them.

And yet, Leibniz presents an opposite vision to that of Newton, according to which space is a substance separate from matter. His relational point of view affirms that without matter it is impossible for a *situation of objects* to exist, and consequently a relational space. There are essential differences between Leibniz and Descartes' thought. We shall discuss some of them in this chapter and later on.

It is important to note that some of Leibniz's ideas were adopted by Ernst Mach in his critiques of absolute motion, such that we can think about his work as the precursor to non-Newtonian mechanics. In addition, the dispute between Leibniz and Newton about the creation of infinitesimal calculus is well-known as one of the most famous of intellectual polemics. Finally, Leibniz made a fundamental contribution to the construction of the conceptual basis of classical mechanics; we will return to his thought on this in the discussion of the concept of force.

In order to finalize this section, it is important to emphasize the transformations of the conceptions of space that occurred in the fifteenth century, by means of the Renaissance artists. New forms of spacial representations changed the basis for new theorizations[14]. In different contexts, regarding the Scientific Revolution, Alexandre Koyré[15] and René

---

[13] Pierre Thuillier states: *It was in Florence, at the beginning of XV century (Quattrocento) that painters and architects have formulated the first theorization about perspective, theorization that would have later a profound repercussion on scientific thought. This fact will provide the projective geometry, as well as has stimulated the arising of space concept a fundamental concept to classical mechanics. The geometrization of space should be understood from a historical point of view, also related to a socio-cultural context very rich* (Thuiller 1994).

[14] Alexandre Koyré characterizes Scientific Revolution of the XVII century in the following form: *Although, I will characterize this revolution by two aspects closely related and still complementary: a) the cosmos destruction and, as consequence, the disappearing, in the science – at least in principle, however not always – of all considerations based on this concept; and b) the space geometrization, that is, the instauration of the dimensional space homogeneous and abstract – in spite of it nowadays could be considered real – of Euclidian geometry in replacing the positional and concrete continuum quite different from physics and astronomy pre-galilean* (Cohen and Westfall 2002).

[15] René Thom states: *I believe, that at the origin of Galilean Revolution there existed only the fact that scientific spirit has found itself in the situation to modelize, because precisely the notion of function, several phenomena, which were not previously modelized with good representation. In my opinion, two factors act combined: on one side, a series of practical needs, I think, mainly the artillery, which encouraged the study of the trajectories of cannon´s balls, of the projectiles, and, of the motion of heavy bodies; on the other hand, the notion of function, which is beginning to germinate in spirit of the investigators and then allowed them to describe, with precision and fidelity, the trajectory of a heavy body. I think, that the formation of this image of function – I say image, because function is here more one image than a concept – it was what originated the great galilaic Scientific Revolution* (Thom 1985).

Thom[16] also confirmed and emphasized the importance of these new forms of spacial representations in the development of Newtonian mechanics (Thom 1985).

2.3 The Development of Time Concept

Some historical and anthropological studies have shown that our conceptions of time have changed significantly throughout history. This fact has been reflected in language and has been demonstrated by the different notions that populations have had about the past, the present and the future. On the other hand, psychological studies focused in knowledge have indicated a strong interaction between the origins of the concept of numbers and the origin of language, as well as how our minds conceive the notion of time.

It seems clear that social forms of organization, the system of creeds and other cultural factors, affect our method of feeling and understanding time. Several authors have affirmed our linear conception of time as something associated with a straight line from the remote past to an infinite future was strongly influenced by Judeo-Christian heritage. In ancient Judaism, there were no cosmic cycles. God created the world starting from nothing, and after 6 days he rested. In addition, cyclic conceptions of time are probably directly influenced by the forms of economic production based in agriculture where products are subject to determined periods.

Independent of how these factors associated with social organization genuinely affect our notions of time in various societies, it is clearly possible to trace, in general, significant changes in our notions of history. And yet, in primitive societies, man had a transitional idea of time, as a flux with its inexorability, mainly associated with birth and death. Some rituals which arose in that time had the objective, in a magical way, of interrupting this continuous flux. Some clues to this type of ritual have appeared from the time of Neanderthal man around 60,000 years earlier.

During the Roman Empire, some interesting conceptions of time arose, and we should consider their contributions to more recent notions. One of them is expressed by Lucretius (94–55 B. C.) in his poem *De rerum natura:*

> *In similar way, the time itself does not exist; but from the existence of things a sense about what happened, what is happen and what will happen appear. We cannot say that some people would be possible to realize the time itself, independent of the motion of things or from it quiet immobility.*

Lucretius also remarked upon the relational aspect of time, which was a modern characteristic for that time, where a strong resistance to change beside veneration

[16] Marcel Conche describing the influence of Epicure on Lucretius affirms: He become concise, he develops, he enriches, he neglects and organizes in different manner Epicure´s thought, by giving a new life…(Conche 1981).

to authority was dominant. He was profoundly influenced by Epicurism and by pre-Socratic Greek philosophers, being too much difficult to identify in his thought the origins of these influences[17].

An important philosophical school with a vision contrary to linear time is Stoicism. In spite of the first Stoics having been Greeks, Stoicism became a very widespread doctrine among Romans. This philosophical school was founded by Xenon of Citrus around 300 B. C. He should not be confused with Xenon of Elea. A cyclic vision of time was a fundamental characteristic of Stoicism. As a consequence, all occurences are obliged to repeat themselves in unfinished cycles. And yet, according to the Stoics, at the end of each cycle, the entire cosmos would be destroyed in a huge conflagration before reappearing again. Inside this chain of fatalities, only human will was free.

As we know, there was no significant scientific progress during the Roman Empire. This situation began to change with the very important contributions to science by Arabic peoples. Around the nineth century, several scientific works and others of applied nature from the Hellenistic period were translated into the Arabic language. One such translation was of the famous book by Ptolemy, the *Almagest.*

At this time, Bagdad in certain ways replaced Alexandria in its role as the center of production and diffusion of knowledge. Greek heritage, along with Iranian and Indian traditions, were being assimilated and transformed by Arabians. This knowledge was transmitted to other Islamic countries, as well as to the south of Italy and to Arabian Spain where Cordoba and Toledo were the main knowledge centers of the twelfth century. During this period, devices for measuring time became necessary.

The Muslims needed people with mathematical knowledge to determine for them through astronomy the moments of the day mandated for prayer and the direction of Mecca. Thus, an appropriate device for measuring time was necessary for this purpose. At that time, the main instrument used was the astrolabe. It calculated fixed star positions with respect to the horizon, the sun and moon and the positions of planets related to stars. However, the more important use of the astrolabe was the determination of the hour of the day or the night through observation of the sun's height although these results are not sufficiently precise. Other instruments were also used to measure time. Water clocks have been in use since the time of Archimedes.

Lack of technical knowledge was the main difficulty in solving the problem of measuring time. This problem would be solved only with the development of the Scientific Revolution, which was just beginning at that time. Water clocks were scarce, expensive and only worked in some European cities.

[17] Viviani, Galileo´s disciple, wrote in 1659: *One day in 1641, during the period I lived with him in the village of Arcetri, I remember that Galileo´s idea that a pendulum could be adapted as watch with weights and springs, replacing the usual instrument...But because he had vision problems and couldn't draw and build models in order to determine what kind of device would better to obtain the desired effect, his son Vincenzo came one day from Florence to Arcetri and Galileo talk to him about the pendulum idea. The discussions were made during several days. Finally he decided to adopt the scheme showed in the drawings and to pass to practice by considering the fact that the difficulties that appear in machines usually are not predicted with simple theorization.* See Drake (1978), p. 153.

In spite of some doubts, the origin of the mechanical clock can be estimated around the end of the thirteenth century. One of the more important impetuses for development of the clock was the need to impose more discipline in monasteries. The religious ceremonies were believed to cultivate virtues and disobeying was punished with severity.

Besides these religious motivations, the increasing expansion of sea trade demanded new methods of navigation and the necessity of improving their precision. Latitude measurements could be done easily by traditional astronomical methods since the time of the Phoenician or even more ancient civilizations. Since the fifteenth century, ships were increasingly crossing the oceans and thus the determination of longitude became a real necessity. There were only two possibilities, the first by means of an astronomical prediction, which was impossible with the existing measurement instruments; the second was by means of a clock on the ship. Only around 1530 was a mechanical clock considered a feasible possibility. However, the clock models of this period based on a flywheel and circular escape motion did not provide accurate measurements.

The fundamental step towards obtaining an appropriate clock appeared with the idea of using a pendulum as the control mechanism. In 1591, Galileo realized the regularity of pendulum oscillations with small displacements. In 1641, he was concerned with the application of such a mechanism and designed a clock pendulum, which was built partially after his death[18].

At this time, Huygens (1629–1695), in the Netherlands, had made several modifications in the clock pendulum to achieve an improved version[19]. Despite this fact, the problem of building a clock pendulum that worked with regularity in ships was an unsolvable problem until 1726. An alternative method was found by Robert Hooke (1635–1703). This method was based on energy accumulated in springs, but further developments made by Huygens and Hooke did not solve completely the problem.

Only with the offering of prizes by governments was the solution of the problem of an embarked clock found. In 1714, the British government offered a prize of roughly 10–20 thousand pounds, depending upon the degree of precision obtained by the designed clock. The prize also required small variations of precision with large variations of temperature. A satisfactory solution was not obtained

---

[18] About Huygens ´contribution to building watches, mainly portable watches: *Before to present this communication, I should say, that there are in modern watches two springs in spiral form; one big drive spring, which move the whole machine and other so small, coupled to equalizer. This last one, also called regulate spiral, that we focus our attention now, because is that the object of Huygens' invention. Later on author affirms: The invention of the regulate spiral is dated of January 20th of 1675.* See Leopold (1982), p. 153.

[19] Michel Paty in (Paty 1988) remarks about the second principle of thermodynamics postulated by Sadi Carnot: *It is not the mathematical-physics, developed then about heat phenomena by Laplace, Poisson, Fourier, that the birth of a new science is determined, but Carnot´s observation according to which motion production in steam machines is due "not to a real caloric consumption, but to its transportation from one hot body to other cold body". The result was the second thermodynamics principle.*

until the middle of the eighteenth century by Harrison from England, Le Roy from France and Berthoud from Switzerland. The solution appeared independently with only a few years of difference. Harrison won the British government prize but further developments followed Le Roy's way in 1766. In the decades of the 1780s and the 1790s, a sufficiently precise model became dominant because of the great impulse towards maritime navigation.

Recently, with scientific developments, new problems have arisen with respect to the concept of time. First of all, the second principle of thermodynamics, established by Sadi Carnot in 1823, has given a sense to time, meaning that some natural processes are not reversible in associating the arrow of time with them. This kind of asymmetry is only theoretically validated by the second principle of thermodynamics because Newton's equations of motion are symmetric with respect to time. There is no difference in Newton's equations if time goes towards the future or towards the past.

Another remarkable fact in our conception of time was the arising of the special theory of relativity in 1905 (Stachel 2001). The experiences which prove the constancy of light velocity to any system of reference lead to the reformulation of classical mechanics laws and the substitution of the concept of absolute time by a time dependent on the system of reference.

In spite of this new perspective for the concept of time, both classical and quantum mechanics still use reversible time. Finally, it is important to emphasize that recent studies on dissipative system dynamics, the chaos theory where certain new concepts such as complexity and others appear, have created a new context with new questions related to time.[20]

2.4 The Concept of Force and Its Controversies

With the success of Newtonian mechanics for solving problems, mainly because of the application of differential and integral calculus using the second law of motion and by means of the Lagrange formalization one century later, the concept of force was established until the appearance of the general theory of relativity in 1916.

The opposition to Newton's concept of force came from the idea of universal attraction that he proposed in *Principia.* Thus, the concept of force was seriously

[20] Prigogine is much too clear to affirm: *We know that Newtonian conception about modern science was under the protection of almighty god, supreme guarantee of a rationality that appeared since the absolute monarchy. In this perspective, that attributes omniscience to a god that dictates the nature laws, the time couldn't be other thing but an illusion. In this sense we can say that our epoch is trying to rediscover the time. But initially it is worth to know, at the beginning of XX century, how we can conceive the relation between the historical time, irreversible, and physical time, reversible. It is interesting to see that there was a kind of consensus. Famous thinkers as Einstein, Bergson or Heidegger, in spite of their differences, they shared the idea that an irreversible time is not the objective of a true science.* See (Forti et al 1996).

contradicted by Cartesians, especially by d'Alembert. The Cartesians did not accept the propagation of forces at a distance, stating that this meant attributing to force a mysterious or metaphysical character to which Newton did not have a reasonable response. To Newton, the problem of completely understanding the concept of force was not solved because of the lack of a cause for producing a force at a distance.[21]

However, the critique of the concept of force had deeper roots than merely the heritage of old cultural elements since ancient times. Aristotle, as we know, did not accept the idea of a void. If the velocity impressed by force is proportional to it and inversely proportional to the resistance imposed by the medium, in a void the resistance would not exist, implying an infinite velocity. This impossibility justifies the impossibility of existence of a void. In addition to the force propagation, it is necessary for a medium to guide the force from the starting point to the point of arrival. Consequently, other conceptions appear to explain the motion of the planets around the sun, as well as several theories about media that fulfill the space. These theories and ideas encouraged studies that followed different directions. Ether's ideas influenced studies in fluid mechanics that also contributed to the building of electromagnetism and modern field theory.[22] The attempts to explain the action of gravity by mechanical processes is an alternative way to use any action at a distance. These theories were developed from the sixteenth to the nineteenth century ever related with the ether conception. René Descartes proposed in 1644 that no empty space can exist and, consequently the space must be filled with matter. According to him every motion is circular, so that the ether is filled with vortices. He also distinguishes different forms and sizes of matter in which heavier matter resists more strongly to circular motion.

Following similar Descartes's premises, Christian Huygens between 1669 and 1690 proposed a more exact vortex theory. His model was the first theory of gravitation which was represented mathematically. He assumed that the ether particles are moving in every direction but being thrown back to the outer borders of the vortex, thus causing a greater concentration of fine matter at the outer border.

---

[22] According to Harman: *While Thomson has developed a theory of a continuum ether as a physical materialization of the full of Faraday´s force, Maxwell has explored the geometrical implications of the force lines, and also has developed a physical model to particles ether to represent a transmission of the field action through contiguous ether particles...Later the author states: The assumption about the elasticity of the magnetic-electric mean has permitted the generalization of the theory to electrostatics, with the elastic stress of the mechanical ether which corresponds to electrostatics field, and hence he has completed field mechanical theory of Maxwell. The theory had an unforeseen implication. Maxwell has calculated the speed of elastic transversal waves which corresponds to a propagation of an electric displacement in the magnetic-electric mean and has found that elastic transversal waves were transmitted with the same speed of the light waves.* He concluded: *light consists of transversal waves within the same mean which is the cause of electrical and magnetic phenomena.* See (Harman 1982), p. 84 and 93.

[21] About this problem Newton wrote as follows: *But, until now, I couldn´t discover the cause of these properties of gravity regarding the phenomena, and I do not invent hypothesis.* See (Cohen and Westfall 2002), p. 312.

In this model the fine matter presses the heavier matter toward the center of the vortex.

With these considerations Huygens also predicts that centrifugal force is equal to the gravitational force which acts in the direction of the center of the vortex. Newton developed at this time his theory of gravitation which is based on attraction. He also discovers that gravity follows the inverse square law and asserts that this is consistent with Kepler's law of motion. However, ether theory was overcome by Einstein's special theory of relativity in 1905, based on experiences that confirmed the constancy of light velocity.

Recent cosmological studies using the Hubble telescope have shown that there exists in the universe a dark matter and energy bringing back the notion of a subtle medium filling intergalactic space. Again, the theory of ether emerges in another scientific context.

The discussions about the concept of force are not restricted to a polemic against the Cartesians. Another important figure to criticize Newton's conceptual framework was Leibniz. (He criticized the Cartesians as well.) As was mentioned earlier, he did not agree with the concept of absolute motion and presented original ideas about the concept of force, as we will see. Leibniz's critique against the Cartesians, especially Descartes, is a very famous scientific dispute. The question refers to the quantity that is conserved during motion. To Descartes, that quantity is the *momentum*, while to Leibniz it is the vis viva, or kinetic energy.[23]

Because of its historical importance d'Alembert's disagreement with Newton's concept of force should be explained. D'Alembert not only criticized the Newtonian concept of force but proposed in his *Treatise of Dynamics*, published in 1743 (D'Alembert 1921), a new approach for solving the problem of a system of particles in motion. This proposal is known as the d'Alembert'principle. This principle has been transformed since its original formulation, starting from the dynamics of the problem of a shock of particles to a method for transforming a dynamic problem into a static representation. It appears in mechanics text-books without historical reference to those transformations (Oliveira 2001).

In his treatise, d'Alembert states: Hence, I have banned completely the forces inherent to bodies in motion, things obscures and metaphysical which are not capable to clarify but to bring darkness to a clear science by itself. He continues by discussing how to estimate forces which act on bodies, affirming: It is only by means of obstacles that are found by the body and by the resistance imposed... Mainly by the obstacles that a body can overcoming or by the resistance that it can opposes... more one can say if the force is big...Although is only in equilibrium or in retarded motion that one could search for measurements (of forces)...Then, in the equilibrium the product of the mass by the velocity, or, that is the same, the quantity of motion, can represents the force.

[23] In *Dynamics Writings*, (Leibniz 1991), p. 3, we find Leibniz´s vision with respect to the question of kinetic energy conservation (vis viva) about the famous debate with Descartes who postulated the constancy of the quantity of motion in the universe.

Obviously d'Alembert's concept of force is different from Newton's conception. The "controversy" is focused in this concept. In some aspects d'Alembert's approach for calculating forces coincides with the calculation of the variation of quantity of motion that is the effect caused by forces. In Newton's *Principia*, the second principle has established that if there are variations of motion, it is due to the presence of forces. Then, the two approaches are similar because Newton used the same model of impact that played a major role in Newton's dynamics, both in development and exposition of the concept of force, as well as providing insight into the third law. In fact, the differences appear because a mathematical representation in a differential form to unify both perspectives is missing. In 1752, Euler proposed that representation.

An important aspect of Newton's concept of force that is presented in *Principia* is the relation between the forces and motions produced by them. Starting from this point, he identifies absolute and relative motions. At the beginning of *Principia*, in *Definitions,* one reads: But we may distinguish rest and motion, absolute and relative, one from the other by their properties, causes, and effects. It is a property of rest that bodies really at rest do rest in respect to one another. And therefore as it is possible, that in the remote regions of the fixed stars, or perhaps far beyond them, there may be some body absolutely at rest; but impossible to know, from the position of bodies to one another in our regions, whether any of these do keep the same position to that remote body, it follows that absolute rest cannot be determined from the position of bodies in our regions. He continues: The causes by which true and relative motions are distinguished, one from the other, are the forces impressed upon bodies to generate motion. True motion is neither generated nor altered, but by some force impressed upon the body moved; but relative motion may be generated or altered without any force impressed upon the body. Newton then describes the famous experience of a rotating vessel to explain relative and absolute motion: If a vessel, hung by a long cord, is so often turned about that the cord is strongly twisted, then filled with water, and held at rest together with the water; thereupon, by the sudden action of another force, it is whirled about the contrary way, and while the cord is untwisting itself, the vessel continues for some time in this motion; the surface of the water will at first be plain, as before the vessel began to move; but after that, the vessel, by gradually communicating its motion to the water, will make it begin sensibly to revolve, and recede by little and little from the middle, and ascend to the sides of the vessel, forming itself into a concave figure (as I have experienced), and the swifter the motion becomes, the higher will the water rise, till at last, performing its revolutions in the same times with the vessel, it becomes relatively at rest in it. Because of the existence of forces acting on the vessel, exerted by the cord in the untwisted motion, there exists absolute motion of water, but the relative motion of it to the vessel could exist or not, as is pointed out by Newton. This experiment has two purposes. The first is to demonstrate the necessity of considering an absolute space as the better method for observing inertial forces acting on fluid mass. The second objective is to contradict the gravitational model proposed by Descartes, according to which the rotation of matter around the sun maintains the planets in their orbits.

Descartes made an analogy with the vortices generated by water flowing around some objects the velocities of which are greater close to them and smaller far away from them. That is what happens with the planets around the sun.

During the period 1680–1690, the same in which Newton worked on his masterpiece, Leibniz developed original ideas about the concept of force and motion. His considerations on force are quite different from Newton's and in some cases in opposition to them. To Newton, forces act on bodies as external attractions or repulsions. To Leibniz, force is a quality inherent to bodies, being an intrinsic property. Another difference between the two mathematicians is that Newton abandoned metaphysical consideration in spite of the fact that his concept of force at a distance does not provide a "scientific" explanation with respect to its propagation. In Leibniz, the explanatory framework of metaphysics comes back to help his construction.

Leibniz uses the concepts of primitive and derivative forces such that each one of these two types can be subdivided into active and passive.[24] Thus, it is possible to have primitive forces, active or passive, and derivative forces, also active or passive. According to Leibniz, primitive forces, active and passive, characterize the body constitution of a given substance. He states: *Primitive force distinguish itself from secondary force (i.e. derivative) which is called moving force is a limitation or an accidental variation of primitive force.* Derivative forces are the forces of more interest to physicists. Leibniz affirms: By derivative forces one denominates that which actually act one on the others or are actuated by others, I understand... only what is connected to motion (local motion) and, by the way, has the tendency to produce later the local motion.

He finalizes his comments as follows: Are these motions (i.e. derivative forces) to which the laws of action and reaction are applied, laws that are understood not only by the reason, but are also corroborated by common sense through the phenomena.[25]

Finally, primitive forces, active and passive, are the substantive parts of matter, while derivative forces, active and passive, are accidental forms or different ways for primitive forces to exist; likewise, the body form is an accident of one thing extended.

2.5 The Concept of Mass

If we analyze the concept of mass before Newton, we realize that it is confounded with the energy concept, obviously the form as to how this concept was understood before Faraday and other physicists of the nineteenth century. In the classical period, Empedocles' classification, according to which any substance is

[24] It is important to note that the concepts of primitive and derivative had been used by Leonardo da Vinci with respect to motion: Primitive motion is that which has the animated thing during the time that is linked to its cause. Derivative motion is that the animated thing has through the air after to be separated to its cause. Derivative motion has its origin in primitive motion and never has the speediness or power equals to speediness or power of this primitive motion. See (da Vinci 2004), p. 72.

[25] See (Leibniz 1991), p. 59.

constituted of some combination of earth, air, fire and water, was dominant for a long period of time.[26]

For the purpose of this book, the concept of mass comes from Newton and is one of the basic concepts of classical mechanics, this being the concept through which the principle of inertia (first law of motion) and the second law of motion are expressed. Obviously its invariability is assumed. Then, it is possible to distinguish mass and weight of a given body. Mass, using such a definition, also provides the definition of universal gravitational law because the force of gravitational attraction is related to the product of the masses involved and inversely to the distance of separation squared.[27]

After Newton, the concept of mass could be associated with chemical thought mainly through the works of Lavoisier as well as the energy concept. In a broad sense, the concept of mass is related to Einstein's equation $E = mc^2$, where mass and energy are interchangeable quantities. In his book *Out of My Later Years* (1950), Einstein presented the derivation of this equation in a concise way. It is important to note the simplicity of this derivation making use of only three previously known laws: (a) The law of the conservation of momentum; (b) The expression for the pressure of radiation, that is, the momentum of a complex of radiation moving in a fixed direction; (c) The expression for the aberration of light (influence of the motion of the earth on the apparent location of the fixed stars).

Lavoisier demonstrated, in the context of physics and chemical phenomena, the law of conservation of mass that one can enunciate as follows: *The total mass of a closed system stays constant for any physical–chemical transformation to which it can be submitted.*[28]

To achieve the above law, Lavoisier performed an experiment in 1768 which consisted of boiling water over a period of 101 days, measuring the weight of the whole system before and after this operation. This was done in a precision balance. It was observed that the whole weight remained constant.

This law permitted great advances in chemical transformation as the law of definite and multiple proportions, enunciated by Proust and Dalton. It expressed the relations among the elements of a chemical compound. And yet, of great importance for chemical purposes, was the periodic classification of elements, assembled by Mendeleyev in 1869. The criteria for classifying the elements is through the association of their properties with their atomic weights organized in a table of growing order. Another important experiment was performed by J. J. Thomson in 1897. He measured the ratio between mass and the charge of the electron in an

---

[26] See *The Fragments of Empedocles*, 1973, pg. 35, translated by William E. Leonard, Open Court Publishing Company, Illinois, USA.

[27] The first quantity defined in Newton´s *Principia* is exactly the mass: The quantity of matter is the measurement of mass, obtained starting from his density and volume. See (Newton 2002), p. 39.

[28] Lavoisier wrote about this question: Indeed, I can consider the substances in presence one of the others and the obtained result, as an algebraic equation; and supposing successively each one of the elements of this unknown equation, one could find a value and to correct the experience by the calculation and the calculation by experience. See (Kahane 1974), p. 130.

experiment performed using discharges of cathode rays which were identified as electrons (Blondel 1994).

After some digression, we come back to the concept of mass as established by Newton. Then, we can also analyze the concept of inertia force as an inherent property of the body, meaning the resistance offered by the body to any change of its state of motionlessness or uniform motion in a straight line. It is worth emphasizing the difference between the inertia principle formulated by Newton and the alternative presented by Descartes based on motion conservation which was conceived in a geometrical perspective. Newton attributes a quantity to the concept of mass as the matter quantity of the body. In the context of Newton's second law, mass appears as a coefficient of proportionality between force and acceleration.

In recent times, starting from experiments performed with cathode rays (electrons), it was possible to detect a mass variation with velocity. The first idea was to associate this variation with a kind of electromagnetic mass in opposition to a mechanical mass without variation. What is observed is that electrons with very high velocities, near to light velocity as result of beta rays, have their total mass increased with this velocity without a constant residue which would be the mechanics mass. This fact provided some opposition to mechanics itself, as the fundamental theory of physics and some physicists tried to replace mechanics with electromagnetism. With the special theory of relativity, mechanics and electromagnetism had been put on a new basis.

With this new theory, mechanics was really refounded. It was also necessary to redefine the concepts of space and time according to the finite value of the velocity of light. The previous definition of mass had lost its meaning and its invariability, and was defined as $m = m_0/\sqrt{1 - (v/c)^2}$ where m is the inertial relativistic mass, and varies with the ratio of the body velocity over light velocity. The value m_0 is known as the motionless mass that is Newtonian mass. According to the expression, to low velocities, both masses are equal. This variation of mass with velocity was demonstrated by Einstein in his third paper published in 1905, entitled: *On the Electrodynamics Bodies in Motion* (Stachel 2001).

As a consequence of this new theory, Einstein had established the equivalence between matter and energy by means of the equation $E = mc^2$, making possible the transformation of mass into energy as follows: $\Delta m = \Delta E/c^2$. This equation indicates that a mass variation is only viable for large amounts of energy. Hence, the possibilities of these transformations only occur in the field of high energies as radioactive transmutations with big liberations of nuclear energy.

Another problem related to mass in a Newtonian sense is the equivalence between gravitational and inertial mass. Using gravitational law $p = mg$ and $f = ma$, implying $f = p$ and $a = g$; then, gravitational mass is equal to inertial mass. This equality, used by classical mechanics, was not given a correct explanation for several years. Through a series of experiments performed by L. Eotvos, during the period 1890–1909, that equality was confirmed. Einstein, in 1907, thought the equivalence was between uniformly accelerated motion and a homogeneous gravitational field. These studies led him to the general theory of relativity, the *equivalence principle* becoming a fundamental principle of nature. As a consequence of Einstein's work between 1911 and 1915, this new theory showed a new structure of the physical world, the space–time determined only by the masses of physical bodies which contain and act as sources of field. Body motion obeys the equations of geometrical nature of the geodesic of space–time.

On 4 July 2012, the CMS (Compact Muon Solenoid) and the ATLAS (A Toroidal LHC Apparatus) experimental teams at LHC (Large Hadron Collider) in Geneva, Switzerland, independently communicated to the world that they each confirmed the discovery of an unknown boson. The experimental data indicates that the behavior of the new particle is consistent with a Higgs boson. This particle has been the subject of a 45 year hunt to explain how matter attains its mass. If finally confirmed, it will be the most important scientific discovery of the century (Fig. 2.5).

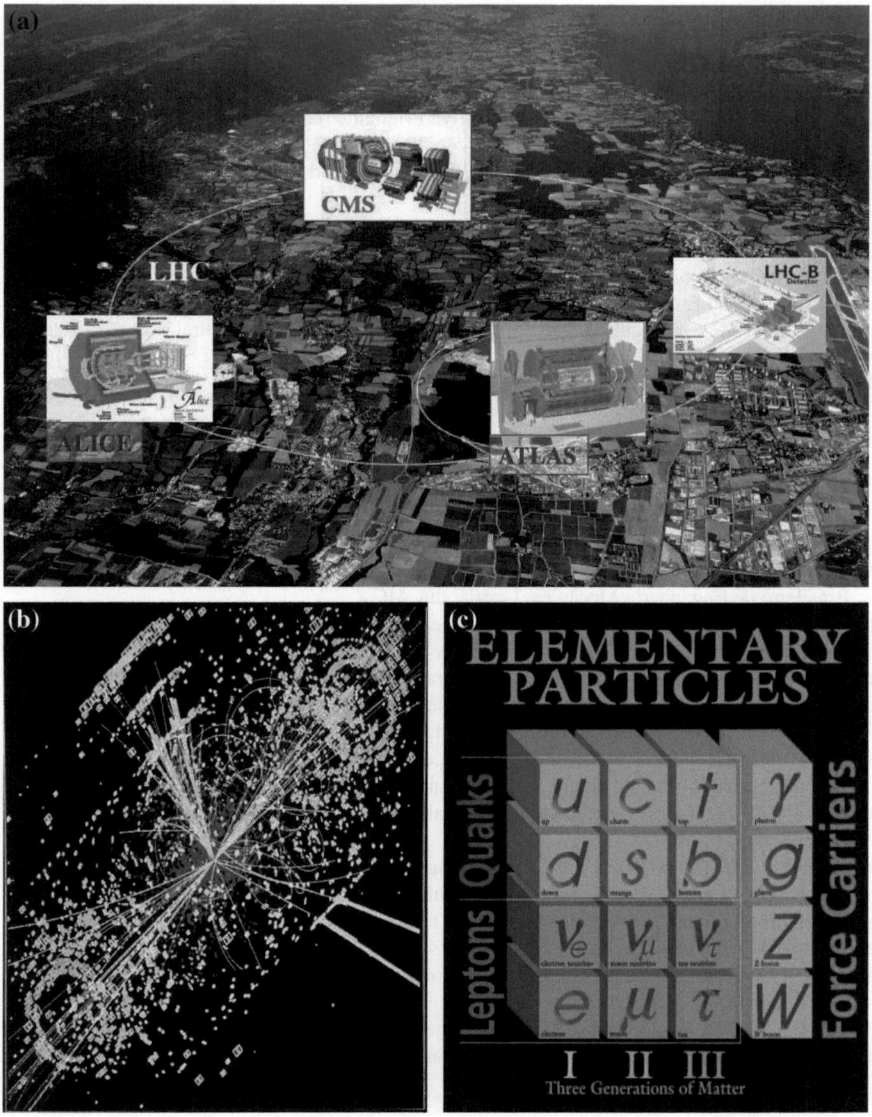

Fig. 2.5 **a** Components of LHC, **b** Traces of Higgs boson, **c** Table of quarks, leptons and bosons

References

Blay M (1992) La Naissance de la Mécanique Analytique, Presses Universitaires de France, Paris
Blondel C (1994) Histoire de l'Électricité, p 116. Cité des Sciences et de l'Industrie, Paris
Cohen IB, Westfall RS (2002) Newton: Textos, Antecedentes, Comentários, p 86. Contraponto Editora LTDA e Editora da Universidade do Estado do Rio de Janeiro, Rio
Conche M (1981) Lucrèce et l'Éxperience. Éditions de Mégare, Paris
D'Alembert JL (1921) Traité de Dinamique, vol 2. Gauthier-Villars et Cie Éditions, Paris
D'Espagnat B, Klein E (1993) Regards sur Matière. Librairie Arthème Fayard, Paris
da Vinci L (2004) Anotações de da Vinci por ele mesmo. Madras Editora, S. Paulo
Descartes R (1991) La Geometrie. Éditions Jacques Gabay, authorized reimpression by Hermann Editors, Paris
Descartes R (1997) Principes de la Philosophie, Lisbon, vol. 9, pp 1–362
Drake S (1978) Galileo at work: his scientific biography. Dover Publication, New York
El Nidditch PH (1995) Desarollo de la Lógica Matemática. Ediciones Cátedra S. A., Madrid
Forti, A. et al (1996) La Mort de Newton, In: Ilya Prigogine (ed.) Irreversibilité et Instabilité des Systèmes Complexes, Maisonneuve et Larose, Paris, p 107
Harman PM (1982) Energy, force and matter: the conceptual development of nineteenth-century physics. Cambridge University Press, UK
Heath TH (1956) The thirteen books of the elements. Dover Publications, New York
Jullien V (1996) Descartes: La Geometrie de 1637. Presses Universitaires de France, Paris
Kahane E (1974) Lavoisier. Pages Choisies, Éditions Sociales, Paris
Leibniz GW (1991) Escritos de Dinâmica. Editorial Tecnos S. A., Madrid
Leopold JH (1982) L'Invention par Christian Huygens du Ressort Spiral Réglant pour les Montres. Vrin, Paris
Mlodinow L (2004) The Euclid's Window. Geração Editorial, S. Paulo
Nagel E et al (1989) Le Théorème de Gödel. Éditions du Seuil, Paris
Newton I (1994) La Méthode des Fluxions et des Suites Infinies, translated to French by M. Buffon. Scientific Library and Technics Albert Blanchard, Paris
Newton I (2002) Principia: mathematical principle of natural philosophy. Book I, Edusp, University of S. Paulo Editors, S. Paulo
Nussenzveig HM (1999) Complexity and chaos. UFRJ/COPEA, Rio de Janeiro
Oliveira ARE (2001) Revisitando o Princípio de d´Alembert. In: COBEM 2001, XVI Congresso Brasileiro de Engenharia Mecânica, Uberlândia, MG
Parkinson GHR (1995) Philosophy and logic. The cambridge companion to Leibniz, pp 199-223. Cambridge University Press, California, EUA
Paty M (1988) A Matéria Roubada. Edusp, Editora da Universidade de S. Paulo, S. Paulo
Paty M (1991) Le Caractère Historique de l′ Adéquation des Mathématiques à la Physique. In: French-Spanish meeting about history of mathematics, Madri
Paty M (1999) La Notion de Grandeur et la Légitimité de la Mathematisation en Physique. Second Journey of Philosophy of Science Jean Largeault, Paris
Petitot J (1990) Idealitées Mathématiques et Realité Objective. In: Seminar of Epistemology of Mathematics, Paris
Prigogine I (1996) Tempo, Caos e as Leis da Natureza. Editora Unesp, S. Paulo
Ross GM (2001) Leibniz. Loyola Editions, S. Paulo
Special Edition of Scientific American dedicated to Borders of Physics (2005)
Stachel J (2001) O Ano Miraculoso de Einstein. Editora UFRJ, Rio de Janeiro
Thom R (1985) Parábolas e Catástrofes: entrvista sobre matemática, ciência e filosofia. Publicações Don Quixote, Lisboa
Thuiller P (1994) De Arquimedes à Einstein. Jorge Zahar Editor, Rio de Janeiro
Truesdell C (1983) Essays in the history of mechanics. Springer-Verlag, Berlin

Chapter 3
The Ideas of Work and Energy in Mechanics

The history of science offers no more surprising case than the phenomenon of simultaneous discovery. We have already named twelve members of the scientific community, each of whom, within a brief period, independently reached the essentials concerning the concepts of energy and its conservation. This list could be extended, in vain, however. This multiplicity already sufficiently suggests that in the two decades prior to 1850 the climate of European scientific thought contained elements capable of guiding receptive scientists to a significant new point of view about nature.

(KUHN, Thomas S. La *Tension Esencial*. México: Fondo de Cultura Editores, 1996)

3.1 General Considerations

It is difficult to trace the history of most physical concepts. This is also true for the concepts of work and energy. One line of investigation has been the process of the invention and construction of machines. In different ages, machines and devices invented to save work have always been researched for a variety of aims. Depending on the age in which the machines, mechanisms and devices were investigated, they have even been seen as associated with magic, due to the sometimes spectacular effects they produced. Since early times, inventors of machines have also realised that the mechanical benefits they offer, have often been accompanied or offset by disadvantages, as a kind of price paid to nature, since she gives nothing for free (Lindsay 1975).

For a long time it has also been known that it is possible to lift a weight through a system of pulleys with much less effort than would be needed to lift it directly. Furthermore, it became evident that the speed of the cable which operated the whole system was much greater than the speed at which the weight was being lifted at the other end of the pulley system. Hence, if the system were operated at low speed, the time required to lift the weight would be excessively long.

A. R. E. Oliveira, *A History of the Work Concept*, History of Mechanism and Machine Science 24, DOI: 10.1007/978-94-007-7705-7_3,

These facts were already known to Heron of Alexandria around 60 AD.[1] This principle, developed throughout history, that the gains obtained by the use of machines were offset by certain disadvantages or losses, contained within it the germ of the concept of energy. In a more subtle way, it was also implicit that inherent to the working of the system, which involved gains and losses or advantages and disadvantages, something must remain constant. And this quantity which remained constant with the system working and which would only be revealed in the middle of the nineteenth century was precisely energy. Evidently we are talking about a system, that is, an idealized machine, in which all its parts comprise rigid elements, its surfaces completely smooth and working without impacts, etc. In reality, as we know, with a working machine part of the energy used in its operation decays in the form of heat or other kinds of loss, such as noise, etc. Nevertheless, the reasoning is valid at the first approach, i.e., discounting losses, the energy remains practically constant. This question of loss immediately raises the question of how to avoid it, that is, how to make a machine more efficient. All these questions are posed historically and the main aim of our investigation is to discuss this development against the background of scientific progress of each epoch. Trying to discover the origins of the energy concept through the study of machines, we can find some clues from antiquity with Aristotle (384–322 BC), who was the first to write about mechanics in his *Physis*, which literally means a study of physics.[2] However, for him, physics had a more general character, something closer to the study of nature. In this treatise, Aristotle presents motion more extensively but does not go into the study of machines. However, there is another treatise attributed to him by some historians entitled *Problemata Mechanica* or *Quaestiones Mechanicae*, which could be considered the first study which tries to explain how machines work. One of the most important characteristics of this study, if we are interested in the origins of the energy concept, is that the approach to machines is more dynamically focused, unlike that of Euclid and Archimedes when they studied states of equilibrium in machines, using the principle of the lever.

According to Pierre Duhem (1997), the author of the above mentioned treatise on the mechanics of machines, regardless of whether it was Aristotle or not, used a basic axiom borrowed from Aristotelian physics, that is, the force applied by one body to another in order to move it was measured by the weight of the body and by the velocity impressed on the same. This means that for the same force applied, the velocity impressed would be inversely proportional to the weight, as if the driving force were the product of the weight times the velocity. These are ideas

[1] In René Dugas (Dugas 1988), p. 32, we read: "Everything indicates that Heron of Alexandria lived during the second century of our era. His treatise *Mechanics* discusses certain simple machines like the lever, the block and pulley, and the screw, separately or in various combinations. The treatise is only available in an Arabian version into which it was translated and published by Carra de Vaux".

[2] Aristotle's works on physics are not contained only in his *Physis*. We also have the treatise on the heavens, meteorology, and on the heavens and the generation of corruption. The study of motion can be found in books ll and lll, (Aristotle 1999) p. 96 and ff.

that already contain germinal notions of the quantity of motion and kinetic energy, and which would only become definitively clearer with the Scientific Revolution or even a little later. The problem of the constancy of a certain magnitude would continue throughout the whole development of mechanics until establishment of the principle of conservation of energy in 1847. The polemics and debates around this problem were intense, involving the most eminent figures in physics and mathematics of the seventeenth and eighteenth centuries.[3,4]

Following the death of Galileo, the problem of conservation in physics became more pronounced. Descartes studied it in connection with the problem of collision between bodies. These studies led to the principle of the conservation of momentum or the quantity of linear motion. He was so impressed with the reach of his discovery, that he proclaimed it as valid for the entire universe. These studies also led him to the conclusion that a measure of force, an influence which produces motion, was exactly the variation of the quantity of motion in the unit of time. Newton was later strongly influenced by these studies of Descartes. We only need to see how Newton proclaimed the second law in *Principia*, attributing to force a variation of motion.[5]

---

[3] D'Alembert dedicates Chapter IV of his Dynamic Treatise, Vol. II, to the principle of the conservation of vis viva. He enunciated it and made a demonstration for diverse particular cases, including the case of fluids. Commenting on the application of this principle by other geometers, as he called them, d'Alembert affirms: "Huygens is the first, that I know of to mention these two principles, and Bernoulli the first to make use of them in order to resolve elegantly and easily diverse problems of dynamics. What I intend to present in this chapter, if not a general demonstration for all cases, are at least sufficient principles to find the demonstration in each particular case", p. 163. And on p. 185: "Daniel Bernoulli, in his excellent work entitled 'Hydrodynamics', presents the laws for the movement of fluids in vessels for the conservation of vis viva, but without a demonstration".

[4] Huygens in (Huygens 1973), p. 48, in his proposal X, affirms: *If a moving body falls vertically, or descends along some surface, and is considered to be carried up again by the impetus along some other curve, it will always have the same velocity at points with the same height when ascending as descending.* In proposition IV, p. 108, we read: *If a pendulum composed of several weights and released from rest were to traverse some part of its complete oscillation, and then its individual weights were imagined, under release from constraint to convert their acquired speeds upward and to ascend as far as they can, when this has occurred, the center of gravity composed of all of them will have returned to the same height that it had before the oscillation began.* From the above citations, we can observe that Huygens took the first steps parting from Galileo's study of falling bodies, in the sense of a balance of potential and kinetic energy, which is not yet an application to machines of the principle of the conservation of the vis viva. He {Huygens} refers his readers to his "Treatise on Equilibrium and the Movement of Fluids", where the principle is demonstrated in a general way see (d´Alembert 1921).

[5] Regarding the influence of Descartes on Newton, Alexandre Koyré sees in the title "*Principia*" an evident reference to "Principles of Philosophy", although in opposition. It emerges to the extent that Descartes' work presents, as the solution for the problem of motion, a mixture of metaphysics and physics. While Newton, when speaking about "Mathematical Principles", tries to benefit from the Galilean revolution, namely, the mathematicization or geometricization of nature. In short, this means mathematical principles in opposition to principles of physics and natural philosophy in opposition to speculative philosophy.

As we already mentioned when presenting the controversies surrounding the concept of force, one of the first polemics involving two great figures in physics and mathematics placed Leibniz and Descartes in opposition. In 1686 Leibniz published in *Acta Eruditorum*, in Leipzig a short article in which he called Descartes' thesis a perversion of mechanics. Leibniz questioned Descartes' thesis on the constancy of the quantity of motion and postulated, as we know, that what remained constant was the vis viva, which was measured by the product of the mass times the velocity of the body squared. Leibniz's argument is based on the following: if we imagine two masses, m and 4 m, and if we release them from a resting position, the first from a height of 4 h and the second from a height of h from the ground, he assures us that each mass, on falling, acquires what he calls the *force* necessary to raise it back to the initial height. This way of visualizing force means work, to use current mechanical terms. Therefore, the same *force* is involved in the fall of the body with mass m as in that with mass 4 m. If we apply the principle of conservation of the quantity of motion to the problem, we see that it is not conserved (Leibniz GW 1991 in Chap. 2).

In the previous chapter we already discussed the most important ideas of Leibniz concerning the concept of force. Besides those and the principle of the conservation of kinetic energy *(vis viva),* we could add another set of theoretical elaborations which also justify considering him one of the precursors of the idea of potential energy. Commenting on the collision between two bodies, Leibniz argued that if they were perfectly hard and inflexible, when they collide with others, they would instantaneously change their directions and velocities. As this was not possible, since it would violate the principle of continuity, such atoms could not exist. In this criticism of the atomistic view, we find justification for existence of the elasticity of bodies. Another line of argument he uses to explain the elastic property of bodies is related to the problem of impact, when he discusses the laws of collision formulated by Huygens and which greatly influenced him. He continued to maintain that in collisions the bodies are stretched and deformed and it is by virtue of their elasticity that they return to their original shapes and, in doing so, move away from each other. It should also be pointed out that Leibniz in his discourse on the physical world turned frequently to metaphysics, which is not Newton`s case. All these considerations lead to the conclusion that Leibniz is not only one of the founders of mechanics but one of the forerunners of analytical mechanics, his successors being d'Alembert, Euler and Lagrange.[6]

Another observation we deem important is that this analytical branch of mechanics, unlike the Newtonian focus which always uses the concept of force, would later be acclaimed as an extremely fertile means for the solution of problems in physics and engineering, known as the *methods of energy.* These methods, well suited for the solution of an enormous array of problems, have as their conceptual basis the kinetic and potential energy of a system or body.

[6] In the next chapter the history of analytical mechanics will be studied in detail.

3.2 The Work Concept and Its Formalization

The concept of work formalized in the eighteenth century in the context of statics must be analysed in other situations and in different epochs, in order to have a correct understanding of its true significance when applied to machines at the beginning of the nineteenth century. This means to say that the ideas which gave rise to the concept of work developed in different directions and in widely differentiated contexts. Therefore, the ideas which arose from the study of the equilibrium of the lever since Aristotle, and passed on by Heron of Alexandria, Archimedes and more recently by Leibniz and others, would give rise to the principle of virtual work (VWP). Another direction and context of the evolution of the ideas of work, concerns the motion of machines in the sense of finding a quantity which remains constant after the system leaves its state of equilibrium. This context is what interests us most, and it is to this that we will dedicate most of our efforts. We will also verify that in the context of the development of machines, the concept of work plays a fundamental role.

There are other directions in which the concept of work also develops such as, for example, the study of human work as physical expenditure or visualizing the human body as a machine. This path leads to the study of physiology and began by measuring the capacity of a worker to polish lenses. These investigations were initiated by Guillaume Amontons (1663–1705), who measured the average pressure exerted by men during the polishing, through the velocity of their action and the effective duration of their working day. He then arrived at a measure of power for human work, as being comparable to that of a horse or a heat machine. Amontons anticipates all the elements which would develop a century later with Coulomb.[7]

Another line of evolution in the concept of work, and which we consider extremely important, is the decay of work inside the machine, which led to the study of thermodynamics. In reality it is an offshoot from the principle line followed by this study. We refer to this line of investigation only in passing and to the extent that it relates directly with our main objective, which is to follow the physical concept of work as a theoretical elaboration in the fields of classical mechanics and mathematics. And then to follow the development of a general theory of machines and finally, the process of incorporating the concept of mechanical work into economic thinking.

Historically, the idea of work as the effect which appears by overcoming resistance was already present at the beginning of the eighteenth century. Its precursor was Amontons, to whom we referred earlier, who tried to establish a way to measure human work. He is surprisingly modern. His name is also associated with one of the first projects to build a steam engine, together with Denis Papin (1647–1712) and Thomas Savery (1650–1715) (Witkowski 2004). They submitted a project for a kind of "fire mill" to the Academy of Science in 1699, but never managed to realize it.

[7] See (Oliveira 2004).

In Amontons' studies, which we can consider as pioneering in the question of work as the action of displacing a force, the dynamic force is measured as a product of a static force (similar to a weight) times the velocity and a time. The product *PVt* which would later reappear in the studies of Coulomb (1736–1806) should be understood as work in the modern sense, since the product *Vt* may be considered a displacement.

Regarding the passage from these concepts to the study of machines, more precisely to the development of the work concept in the context of the search for a quantity which is conserved by a machine in motion, we have to consider as a starting point a note written by Antoine Parent (1666–1719) in which he tries to estimate the optimum velocity at which the wheel of a mill should turn in order to obtain *the greatest possible effect.*[8] Efforts in this direction continued with Henri Pitot (1695–1771),[9] in 1725 and with Bernard Forest de Bélidor in his *Architecture Hydraulique*, in 1737–1739, which would become the most popular reference manual among the engineers of the eighteenth century. This manual also has considerable historical importance within the context on which we are focusing, since it would be updated by Navier in the early decades of the nineteenth century and this republication of Bélidor by Navier is one of the fundamental links in the chain which connects the undertaking of Lazare Carnot with polytechnic engineers. In the final chapter we will deal with this problem in detail.

Returning to the trajectory described by the concept of work strictly from the viewpoint of physics, the writings of Jean Bernoulli (1667–1748), deserve special attention, in particular the letter addressed to Pierre Varignon (1654–1722), dated 26th January 1717, in which he makes some generalisations which would later become *The VWP*. Let us look at a section of this letter:

> Let us suppose that various different forces are acting along different lines or directions with a tendency to maintain in equilibrium a point, a line, a surface or a body, and let us also suppose that we impress on the system as a whole a small displacement, always parallel to itself in any direction, or around any fixed point; it is easy to see that due to this displacement each one of the forces will advance or retrocede in its own direction, unless

[8] Jean-Pierre Séris sums up Parent's contribution as follows: "While in the machine in equilibrium, the velocity of the fluid (squared) is taken absolutely as a component of the effort on the area, it is the difference in the velocity of the fluid–velocity of the area which, squared, intervenes in the calculation of the current's effort on the area in motion. This principle, as we shall see, does not introduce any new concept, but adopts an unprecedented procedure". See (Séris 1987) p. 290.

[9] On p. 296, Seris (Séris 1987), affirms: "In reality, Henri Pitot (1695–1771) who in his paper of 1725 entitled '*New Method of Knowing and Determining the Effort for all Kinds of Machines Moved by a Current or Fall of Water*', would present things in this clearer more coherent manner. Knowing that he must obtain the maximum quantity (V–v), Pitot initially 'calculated the velocity that the blades of a machine would have to reach in order to produce the maximum possible effect'. This new presentation, more intelligible and more intelligent than Parent's idea, should not be underestimated. This is what would later popularise (the idea) allowing it to reach a wider informed public. (Euler, Mc Laurin, Daniel Bernoulli), men involved in art (Bélidor) and experimentalists (Deparcieux, Smeaton)".

some of them have their directions perpendicular to the direction of the small displacement; in this case, the force or forces will neither advance nor retrocede: for the advances or retrocessions which we call "virtual velocities", the quantities of each line or tendency will increase or decrease owing to the small displacement; and these increments or decrements are found by tracing a perpendicular from the end of each line of tendency which will be cut by the line of tendency of each force around the position to which it has been shifted by the small displacement, a small portion of which will be the measure of the "virtual velocity" of this force.

At the end of this long citation, Jean Bernoulli draws an explanatory diagram which makes the operation he performed quite clear. The letter ends with a General Proposition or Theorem XL which synthesizes his ideas:

In every case of equilibrium of forces, in whatever manner they are applied, and in whatever direction they act on the others, mediately or immediately, the sum of the positive energies will be equal to the sum of the negative energies, taken as positive.

This is the founding text of the VWPs. Still in the field of physics, the work of Daniel Bernoulli (1700–1782) entitled *Hydrodynamics,* of 1738, returns to the question of work in the context of *the principle of vis viva forces*. This text is very important as a starting point for a series of studies as much theoretical as machine linked, as we shall see during the course of this investigation. It is also important from the terminological point of view, since it introduces the term labour into the concept of work.[10]

Finally, in 1798, Fourier published an article entitled "Report on Static and the Theory of Moments" (Fourier 1789). In this study, he starts with Jean Bernoulli and takes mathematical development and the generalisation of the physical concept of work, applying it from a material point all the way up to a complex system, in the form we currently know, even though the term work had not yet been adopted. It should be pointed out that Fourier`s report is one of the most complete studies on the principle of virtual velocities. In short, at the end of the eighteenth century the concept of work was perfectly established and formalised, and could

[10] Daniel Bernoulli, in (Fourier 1789), dedicated all of Chap. 9 to the application of his theory of fluids to machines. The title of this chapter is quite suggestive: *Concerning the Movement of Fluids Which are not Impelled by Their Own Weight, but by an External Force, Particularly Hydraulic Machines and their Highest Degree of Perfection Attainable, and How this can be Perfected Later Through the Mechanics of Solids as well as that of Fluids*. This chapter is divided into three parts, namely: Part One–Concerning hydraulic machines expelling water upwards without appreciable impetus. Part Two–Concerning hydraulic machines transporting water without appreciable impetus from a lower to higher position. It is precisely in this section, that on p. 202, his famous mechanical principle (known as) Rule 10, explicitly appears as a mechanical principle, which he applies to fluids; and which he later applies to the case of Archimedes' screw. Part Three - Concerning machines which are moved by the impact of a fluid, in the same way as the force of the wind. This title refers, evidently, to a water wheel, in which the fluid communicates its free-fall motion to the blades of the wheel, causing it to turn. Bernoulli considered that if the fluid had a velocity V, we must consider the relative velocity $(v–V)$ as driving the machine. He also made several considerations on the motion of the blade, placing an algebraic equation between v and V, whose maximum is $V = 1/3\ v$.

be used in the study of machines. Lazare Carnot (1753–1823) would be the first to create a general theory of machines, although he uses different terms to designate work. Other works also deserve attention for their approach to machines. In this sense we should mention John Smeaton (1724–1792) in England and Jean Charles Borda in France. Smeaton was an instrument maker, whose name is associated with the construction of more efficient models of the Newcomen machine, improving its manufacturing process. However, it was nothing comparable to the great theoretical efforts of Carnot.

So the foundations were formed and conditions created for him to develop the first general theory of machines integrated into the theoretical framework of rational mechanics, as we shall see more profoundly in Chaps. 5 and 6.

As we have commented throughout this text, the creation of this theory would be the work of Lazare Carnot. It was made public for the first time in 1780, but would only be completed in 1803. His work occupied an intermediary position between rational mechanics and a more applied science typically used by engineers. His general theory is also an important starting point for a series of developments and applications carried out by a group of engineers from the *Polytechnic School* in Paris who are responsible in great part for the applied mechanics which arose in the early decade of the nineteenth century, mainly from the viewpoint of the constitution of this discipline.

The only reason that Lazare Carnot`s undertaking did not have wider repercussions was because he is largely obscured by the brilliance of Lagrange's *Mécanique Analytique*, published in 1788. In any case, at the beginning of the nineteenth century the theory of machines developed strongly in France where it was incorporated into the academic teaching of engineering in the *Polytechnic School.*

As will also be seen, the concept of work followed the course and evolutionary line of the latest developments in the field of machine mechanics and holds an important position in Carnot's theory.

3.3 The Development of Virtual Works Principle

The known VWP has been intensively used as an alternative method to the equations of equilibrium taken from Newton's laws for a given mechanical system. If these systems have the particularity of possessing various degrees of freedom and are constituted by interconnected sub-systems, the VWP has a series of advantages over the conventional method of equilibrium, since it operates only with those forces (external) which do the work, thereby avoiding waste of time with the manipulation of internal forces, whose work total, since they appear in pairs with crossed signals, is null. Another advantage of VWP compared with the traditional method is that it permits the study of the nature of equilibrium. The position of equilibrium of a system being associated with the problem of finding the maximums and minimums of its potential function, the VWP is directly related to the stability of the system, which occurs when this function becomes a minimum. In the inverse case instability occurs, evidently the system may still present a type of equilibrium called *indifferent*, when the first and second derivatives of the potential function annul each other simultaneously. An important observation must be made. In older texts, this principle is always referred to as the principle of virtual velocities, this is due to the fact that the term work was only adopted after 1829 by Coriolis, Poncelet and the group of polytechnic engineers. In this way the concept became part of vocabulary of mechanics and from then on the term work began to replace the term velocity in the famous principle. There follows a chronology of the most important investigations into the VWP after Jean Bernoulli`s above mentioned letter to Varignon:

(a) In 1788, *Essai sur les machines en géneral*, by Lazare Carnot.
(b) In 1796, *Memória sul principio della velocita virtuali*, by Fossombroni.
(c) In 1797, *Théorie des fonctions analytiques*, first edition, by Lagrange.
(d) In 1797, *Considérations sur le principe des vitesses virtuelles*, text by Poinsot which was not published.
(e) In 1798 we have three of the most important works. The first is *Memoire sur la statique*, by Fourier. The second is *Sur le príncipe des vitesses virtuelles*, by Lagrange and finally the third *Sur le príncipe des vitesses virtuelles et la décomposition des mouvements circulaires*, by Prony.
(f) In 1799 the theme of virtual works is dealt with in *Mécanique Céleste*, by Laplace and the problem is approached in a second study by Prony entitled *Mécanique philosophique*.
(g) In 1803, Lazare Carnot returns to the matter in his *Príncipes*.
(h) In 1806 we have two interesting studies. The first called *Théorie general de l'équilibre et du mouvement des systèmes*, by Poinsot and the second is *Démonstration génerale du príncipe des vitesses virtuelles, degagée de consideration des infiniment petits*, by Ampère.
(i) In 1811 we have the second edition of *Mécanique analytique*, by Lagrange.
(j) In 1813 we have the second edition of *Théorie des fonctions analytiques*, by Lagrange.

Some observations must be made. Firstly, the period beginning with Lazare Carnot in 1788 and ending with the death of Lagrange in 1813, was extremely prolific in terms of studies on the VWP. We should point out that the great majority of these investigations are part of the fundamental studies of mechanics itself as a scientific discipline and the VWP is practically incorporated into its conceptual base as a fundamental principle and foundation of mechanics. This at least is the case with the *Principles* by Carnot, *Mécanique Analytique*, by Lagrange and *Mécanique Céleste* by Laplace. The most important isolated texts which deal specifically with virtual works are those by Fourier, those by Lagrange, which also deal with it in the context of analytical mechanics, as well as those by Prony and Poinsot. The texts by Carnot in which the VWP appears are of great historical interest for our investigation and for this reason will be studied in a specific item. Carnot presents the concept, which he elaborated, of geometric motions which are closely related with the VWP, as will be shown later. Let us see then in summary what the contributions of Fourier, Lagrange, Prony[11] and Poinsot deal with.

3.3.1 Fourier (1798)

The full title of Fourier's work is *Memoire sur la Statique Contenant la Démonstration du Príncipe des Vitesses Virtuelles et la Théorie des Moments*. In a general way, the demonstration of the VWP in this study is based on the principle of equilibrium of the lever. Fourier also borrowed from Galileo the notion of moment, as a product of the force times the virtual velocity of the point at which it is applied. Some imprecisions may be observed, as for example, the use of a sign contrary to that which is normally used nowadays, which is explained by certain indefinitions in the terrain of the theory itself, both in the differential calculus and in the concept of energy. Fourier initially considers the case of two equal opposite forces acting on an inflexible line, and from the analysis of the connection established between the two rigid surfaces, he proves the VWP for the case of a rigid body. The other cases analysed are those of flexible bodies and systems containing incompressible fluids.

In the demonstration presented by Fourier which, according to the historian René Dugas (1988), is *the logical reduction from the theorem of virtual works to the principle of the lever,* he also shows that this principle and the principle of the composition of forces come down to the same thing. In other epochs, some investigations had already reached the principle of the lever from the parallelogram of forces. Fourier, in the study mentioned, took the inverse route. In going deeper, he was searching to give a better foundation to the VWP based on the principle of the lever, which had already been done satisfactorily by Archimedes and Huygens (Heath 2002).[12]

3.3.2 Lagrange (1798)

Lagrange's demonstration of the VWP appeared for the first time in the 5th Section of the Journal of the *École Polytechnique* and was republished with some modifications

---

[11] For a consistent study on the history of the VWP, see (Poinsot 1975). However, this author commits a mistake when he attributes adoption of the term "work" to Poncelet, instead of to Coriolis.

[12] Archimedes, in his "Concerning the Equilibrium of Planes" or "The Centres of Gravity of Planes", Book I at (Poinsot 1975) uses the principle of the lever to study the gravity of flat figures.

in the 2nd edition of his *Mécanique Analytique* in 1813. His demonstration is quite succinct containing only four pages.[13]

Right away in the introductory paragraph, Lagrange mentions the same kind of problem faced by Fourier: *In the demonstration we made of the principle of virtual velocities, the same was made to depend on the principle of the composition of forces or of the equilibrium of the lever.* What he questions is the fact that these two principles, even being fundamental principles of static, require demonstration. Therefore, in the search for a more elementary basic principle, Lagrange turns to the principle of the equilibrium of a lifting block (pulley), understanding it to be self evident and self explanatory and leading directly to the principle of virtual velocities. In the second edition of his Mécanique Analytique, Lagrange calls this principle the principle of pulleys.

The procedure employed by Lagrange consists of making the representation of the forces applied by a mechanical system comprising simple machines. This made his demonstration similar to that of Fourier with the difference that the levers were replaced by a combination of pulleys.

Everything indicates that the system set up by Lagrange was inspired by the work of Stevin, since the latter actually announced the VWP in a situation restricted to a system of pulleys in his famous *Hypomnemata Mathematica.*[14]

Furthermore, Lagrange cites Stevin (1548–1620) on the first pages of his most well known book.

3.3.3 Prony (1798)

Prony's study entitled: *Sur le Príncipe des Vitesses Virtuelles et la Décomposition des Mouvements Ciculaires*, is divided into two parts, as the title itself suggests.

Initially he considers the case of rigid systems and defines five elementary motions, from which it is possible to compose any motion. He then calculates, using the parameters representing the five motions already defined, the elementary displacement of any motion, from any point of the animated solid in motion. He then obtains the projection of the virtual displacement in the direction of the

---

[13] In the second section of his "Mécanique Analytique", p. 12, (Lagrange 1989), we read: "The general law of equilibrium in machines is that the forces or potencies are between themselves, reciprocally like the velocity of the points at which they are applied, estimated according to the direction of these potencies. It is this law which contains what we commonly call the "Principle of Virtual Velocities". This was recognised after a long time as the fundamental Principle of Equilibrium, already shown in the previous section, and which may consequently be considered a kind of axiom of mechanics".

[14] In 1608, Stevin assembled several works which he called "*Hypomnemata Mathematica*". They were translated into French in 1634. According to René Dugas, (Dugas 1988), Stevin's static is developed geometrically in a very similar way to that used by Archimedes. With regard to the Principle of Virtual Veocities, we can find it in Volume IV of his "*Hypomnemata*", where he deals with the equilibrium of a system of pulleys.

force applied to the point. Then, as he already demonstrated in a previous study, from the composition of the forces, the six equations of equilibrium of a solid are applied here to deduce the principle of the virtual velocities of these six equations.

Prony is not entirely satisfied with his demonstration and affirms: *It is necessary, in order to conserve the character of the principle in the virtual velocities equation, to deduce it from a mechanical theorem even more elementary and closer to the true definitions.* In this way he seeks to support himself on *the composition of the forces applied to a single point and on the parallel forces.* The demonstration of the VWP comprises then in demonstrating its validity for different cases, such as that of a single point, an inflexible line or a solid triangle, applying the lever principle to all of them. His method recalls that of Fourier presented earlier.

In the final part of the study, Prony applies the VWP to the case of deformable bodies. However, he limits himself to the case of funicular polygons. Prony's study directly influenced his student Poinsot, who also made an important contribution to the VWP.

The advancements made by Fourier, Lagrange and Prony demonstrate that the VWP is a fundamental principle of mechanics, being directly related to the equilibrium of the lever and the composition of forces. The triggering of many studies on the VWP soon after the publication of Lagrange's *Mécanique Analytique* also indicates that it provoked intense debate in the scientific world. This problem, which had been approached earlier, was one of the preferred matters since the founding of the *École Polytechnique* a few years before.

Other contributions, seeking a convincing demonstration of the VWP, appeared in the following years, and Lagrange himself returned to the theme in the re-publications of his *Mécanique Analytique* and in *Théorie des Fonctions Analytiques*.

3.3.4 Poinsot (1806)

Louis Poinsot, student of Monge, Lagrange and Prony, made a great contribution to mechanics, not so much for having presented a new demonstration of the principle of virtual velocities, but by the fact of having placed the debate about the fundamental principles of mechanics on a new basis. If we bear in mind that the first years since the founding of the Polytechnic School and the first two decades of the nineteenth century, more or less up to the death of Lagrange in 1813, was a very fecund period in terms of discussions about on which basis and on which really fundamental principles mechanics rests, this contribution has a special merit.

According to the great majority of mathematicians and engineers of the time, the principle of virtual velocities was undoubtedly one of them. Lagrange and many others tried to reduce this to the principle of the lever. This is why in his *Mécanique Analytique,* of 1788 there is no other demonstration of the principle of virtual velocities. And, as we know, his mechanics rests in great measure on this principle.

The merit of Poinsot's studies was to provoke a complete reversal of what seemed to be clear, obvious and fully established. His questioning of the principle of virtual velocities was similar to the criticism that d'Alembert made about the concept of Newtonian force, as something obscure and incomplete. He argued the opposite, that the principle in question should be deduced from the general conditions of the equilibrium of the system.

Poinsot then returned to the re-elaboration of a mechanical system and of the relationships between its parts, generalising the problem from the equilibrium of a point to a system of material points. It was due to these criticisms that Lagrange made important modifications regarding the principle of virtual velocities in the second edition of two of his fundamental works: Analytical Mechanics and the Theory of Analytical Functions, published respectively in 1811 and 1813.

3.4 The Least Action Principle

One of the aims most sought after by physics along the years has been to find a principle, the simplest possible, into which all natural phenomena would fit, and which would also allow the calculation of all past occurrences and principally future occurrences. Evidently, this is far from being reached and quite probably does not even exist. Nevertheless, an approximation to this ideal is always possible and the history of physics shows that some results in this direction have been achieved.

Among the more general laws which fall within the appointed perspective is the principle of least action (LAP). In all branches of science to which it has been applied, it is possible to obtain a general explanation of certain characteristics of the phenomena involved, as well as providing the rules which indicate how these phenomena vary in time and space and how to determine them. In this way, the LAP has occupied a central position in modern physics, together with the principle of the conservation of energy. It is worth pointing out that the latter holds a position similar to the former and governs a large number of physical processes.

The principle of the conservation of energy can be deduced from the LAP and consequently is contained within it. The reciprocal, however, is not true, which confers a more general character to the LAP. As an example we can consider the motion of a particle free from the action of any force. According to the principle of the conservation of energy, the particle should move at a constant velocity, but nothing is said about the direction of the velocity vector, since the kinetic energy does not depend on that direction. In principle, the trajectory of the particle could be as much rectilinear as curvilinear, if the motion is considered only from the energy point of view. If we apply the LAP, we conclude that the particle must move in a straight line.

This problem is entirely general and can be generalised. In the case of a spherical pendulum, that is, a concentrated mass moving without friction over a fixed spherical surface, the principle of the conservation of energy only states that during the upward movement, the kinetic energy diminishes by a certain quantity and increases during the downward motion. The trajectory of the particle cannot be determined by the conservation of energy, while the LAP completely resolves all the questions related to this type of motion.

The fundamental difference in the application of the two principles to any problem is that the principle of the conservation of energy provides only one equation, while it is necessary to obtain as many equations as there are variables or degrees of freedom, in order to completely determine the motion of the system. Thus, in the case of the free particle, three equations are needed and in the case of a spherical pendulum only two. Whereas the LAP provides, in all cases, as many equations as there are variables.

We should bear in mind that the LAP provides various equations from a single formula, since it deals with a variational principle. Its great advantage is that from an infinite number of imaginable virtual motions, under the particular given conditions, it indicates a well-defined motion by means of a simple criterion and points to the effective motion. This criterion means that for an arbitrary motion infinitely close to the effective motion, or more precisely, for all infinitely small variations of the given motion, consistent with the given conditions of the connection, a certain characteristic function of the variable is annulled. In this way, an equation is deduced from every independent variable, that is, from every degree of freedom of the system, as a problem of maximum or minimum.

The origins of the LAP go back to Leibniz.[15] In a theorem enunciated by him, it was established that of all the worlds which could be created, the effective world is that which contains, along with all the inevitable bad, the maximum good. Although postulated in the moral terrain, the theorem proposes a variational type solution like the LAP. According to some historians, Leibniz first formulated this principle in a letter dated 1707, whose original had been lost. Later, the LAP would be developed by Pierre-Louis Moreau de Maupertuis (1698–1759), Euler and Lagrange (de Maupertuis 1744).[16]

The LAP is indissolubly linked to the name of Maupertuis, who not only recognized the existence and significance of this principle, but also used all his influence in the scientific world to work for its acceptance. We cannot forget that he was nominated by Frederick the Great as President of the Prussian Academy of Sciences, a position he held from 1746 to 1759.

Maupertuis was born in Saint Malo, France, in 1698, and was one of the great disseminators in France of Newtonian theories, especially that of mechanics and the law of universal gravitation. To this end, he conducted an expedition to Lapland, in 1736–37, to measure the acceleration of gravity, with a view to

---

[15] In March of 1751, Koenig, professor at La Haye, and a member of the Berlin Academy, asserted that Leibniz was the first to formulate the principle of least action. This affirmation was based on a letter from Leibniz to Hermann dated 16th October 1707, whose autheticity was questioned by several scholars of the matter. See (Gueroult 1967).

[16] Although this correspondence has never been found, Leibniz pronounced several times on the principle of least action, it is not improbable that he himself was the first to formulate this principle. Maupertuis' defenders, however, argue that he was responsible for the thoroughness and precision of the formulation. See (Gueroult 1967).

confirming Newton's supposition about the flattening of the earth. The positive results obtained by the expedition strongly influenced the acceptance of Newton's ideas on the European continent.

The LAP was discussed by Maupertius in his paper entitled *Accord des Différentes Lois de la Nature qui Avaient Jusqu'ici Paru Incompatibles* presented to the Academy of Science of Paris in 1744. As we know, this principle was originally formulated for geometric optics and was only later extended to the domain of mechanics.[17]

In his paper, Maupertuis enunciates the three laws which light must obey:

(1) In a uniform medium light moves in a straight line.
(2) When light encounters a medium it cannot penetrate, it is reflected and the angle of refraction is equal to the angle of incidence.
(3) When light passes from one transparent medium to another, its trajectory, after meeting the new medium, forms an angle with the previous trajectory, so that the sine of the refraction angle is always in the same ratio as the sine of the incidence angle.

The so-called law of sines for refraction had already been proposed by Descartes and independently by Snell.

In "Dioptrics" of 1737, Descartes presents a deduction from this law based on mechanical arguments and on the supposition that light moves more easily in a dense medium. Pierre Fermat (1602–1665) disagreed with Descartes' arguments and only with experiments by Leon Foucault (1819–1868) and Hippolyte Fizeau (1819–1896), in the middle of the century XIXth Century would Fermat's hypotheses be confirmed.

Let us see then how Maupertius formulated the LAP, extracting from the famous paper the part in which the principle appears:

> After meditating deeply on this topic, it occurred to me that light, upon passing from one medium to another, has to make a choice, whether to follow the path of shortest distance (the straight line) or the path of least time. But why should it prefer time over space? Light cannot travel both paths at once, yet how does it decide to take one path over another? Rather than taking either of these paths per se, light takes the path that offers a real advantage: light takes the path that minimizes its action. Now I have to define what I mean by "action". When a material body is transported from one point to another, it involves an action that depends on the speed of the body and on the distance it travels. However, the action is neither the speed nor the distance taken separately; rather, it is proportional to the sum of the distances travelled each multiplied by the speed at which they were travelled. Hence, the action increases linearly with the speed of the body and with the distance travelled. This action is the true expense of Nature, which she manages to make as small as possible in the motion of light.

Maupertuis arrived in this way at a problem of minimization. When making this function minimum, he concluded that the sines of the angle of incidence and

[17] For a detailed exposition of Maupertuis' principle, as well as the paper cited, see Suzanne Bachelard at (Actes de La Journée Maupertuis 1975), p. 99.

refraction equals the inverse ratio of the speeds at which light moves in each medium. This result disagrees with that of Fermat and is in agreement with Descartes. This formulation proposed by Maupertuis does not allow conclusions to be drawn regarding the laws which govern the phenomena nor the conditions which must be met.

Following the path opened by Maupertius, Euler[18] extends the ideas underlying the LAP and proposed a principle formulated as a mathematical theorem applied to mechanics for the case of constant energy. Euler affirms: *Since all natural processes obey certain laws of maximum or minimum, there is no doubt that the curves described by the bodies under the influences of arbitrary forces, also possess some property of maximum and minimum.* He further added that the form proposed by his theorem was only applied when the forces depend on the position and that dissipative systems would not lend themselves to a description of this nature.

After Euler, it was Lagrange[19] who expressed the LAP in an entirely general formulation as a principle of stationary action for a general system of n bodies interacting among themselves. This undertaking is made in his *Mécanique analytique,* published as we know in 1788.

In 1834, William Rowan Hamilton (1805–1865) showed that the LAP admits other representations. In this way he established a strong analogy between mechanics and optics, relating Fermat's principle to that of Maupertuis. This was how he arrived at the current and most used form of the LAP as a variational principle of mechanics, known as the Hamilton principle.

Initially, the postulation of the LAP did not provoke any major commotion nor did it have any considerable effect on the advancement of science, even after the general formulation made by Lagrange. The existence of this principle was considered more a mathematical curiosity and even an unnecessary corollary to the laws of Newton. Nor was there any lack of voices raised against its usefulness. In 1837, Poisson called the LAP the *worthless rule*.

Only with investigations by William Thomson (1824–1907), Peter Guthrie Tait (1831–1901), Gustav Robert Kirchoff (1824–1887), Von Neumann (1832–1925) and Ludwig Boltzmann (1844–1906), principally, was it proved that the LPA was the most suitable method for resolving problems of hydrodynamics and elasticity. Furthermore, the LAP was unbeatable when the usual methods of mechanics functioned only with difficulty or even failed. It was only then that its worth and reach began to be perceived. This is clear in the following affirmation by Thomas Tait, made in 1867:

> Maupertuis's celebrated LAP has been, to date, looked upon more as a curiosity and a property of some strange forms of motion than as a useful rule in kinetic investigations. We are strongly impressed and convinced of a much more profound significance attached to it, not only in the abstract dynamic, but also in the theory of the various branches of physical science now beginning to receive a dynamic explanation (Planck 1993).

[18] According to Moreira (Moreira 1998), Lanczos in his "The Variational Principles of Mechanics" affirms that Euler discovered the Principle of Least Action independently of Maupertuis.

[19] As we know, Lagrange in his "Mécanique Analytique", postulated a general formulation of the Principle of Least Action for a system of interacting n bodies. See (Lagrange 1989).

As time went by and with the success of its applications, there was a perception of the fundamental significance of LAP as a general principle which could be applied to systems whose internal physical mechanisms were entirely unknown, or so complicated that they could not be represented by means of ordinary systems of coordinates. Another important fact is that after Boltzmann, Rudolf Clausius (1822–1888) perceived the close relationships between the LAP and the second law of thermodynamics. Furthermore, Hermann von Helmholtz (1821–1894), for the first time, demonstrated the existence of a complete systematic application of the LAP to the three great branches of physics: mechanics, electrodynamics and thermodynamics. This gave the LAP a higher status and deepened comprehension of it as a general principle.

In this way, the LAP followed a course similar to the principle of the conservation of energy which, as we know, was originally considered a mechanical principle, despite its general validity.

More recently, without making use of any mechanical hypothesis, Joseph Larmour (1857–1942) in 1900 and Karl Schwarzchild (1873–1916) in 1903, among others, deduced the fundamental equations of electrodynamics and of the theory of the electron from the Hamilton principle.

The PLA also demonstrated its validity and explicative potential, with regard to quantum and relativistic mechanics. Initially, the probabilistic characteristic of quantum mechanics appeared to exclude the LAP from its field of application. However, Feynman, in a classic study, showed that it was possible to include it in this field and created a variational principle embodying the quantum phenomena.

Finally, one of the most brilliant results attained by the LAP was the fact that Einstein's theory of relativity had shown that it occupied an outstanding position among the laws of physics. This is because the function denominated *action*, according to Hamilton, though not that of Maupertius, is an invariant with regard to all the transformations of Hendrik Lorenz (1853–1928), signifying that it is independent of the observers' system of reference. From this brief historical outline of the LAP, everything indicates that it appears to govern all of nature's reversible processes. What remains open is that it offers no explanation for the phenomenon of irreversibility. According to the LAP all phenomena are able to move in any direction of time, as in the example of Newton's law, able to travel both forward and backward.

3.5 The Energy Conservation Principle

The word energy is reasonably new and its current sense is related to the principle of its conservation, established in 1847.[20] Not that an intuitive notion had not existed for some time, as we saw earlier. Atmospheric discharges, the force of the wind and rain, etc. were well known. However, these powers, as these forces of nature were called, were each considered distinct and unconnected. As we shall

[20] According to Atkins, Ref. (Atkins 2003), p. 83, we owe the term "energy" to Thomas Young (1773-1829). While Robert Locqueneux, affirms that William Thomson introduced the term "energy", this was in 1850. See (Locqueneux and la 1996).

see, the principle of the conservation of energy would not only connect, but also unify them and fit them into the same manifestation of nature.

One of the key persons on the road to unification of the diverse views and manifestations of energy is undoubtedly Michael Faraday (1791–1867), who from an apprentice bookbinder, went on to become an extremely important figure in 19th century physics. Coming from the poorest social classes of English society, he managed to get a job as an apprentice bookbinder which had a singular advantage, *There were many books there*, as he revealed to a friend years later. Sometimes he spent the night alone reading by candlelight those books which most interested him. When he was twenty years old, a visitor to the workshop offered him a ticket for a series of conferences at the Royal Institution, The conferences were on electricity and the occult energies which must exist in nature. These not only awoke Faraday's interest but also pointed the way to a better life. But how was he to enter academic life, if he had not even attended what we now call secondary school? Furthermore, his material resources were practically nul. His father was a blacksmith and had never managed to give him anything.

Faraday then made use of the resource which he best dominated, the art of the bookbinder. He gathered the notes on conferences then drew and inserted some designs of the equipment used in the demonstrations, using leather and engraving tools he bound all the material and sent it to Humphry Davy. After this Davy expressed a desire to know him, when they finally met Faraday was hired as a laboratory assistant. It was to take some years before he and Davy established a natural professional relationship.

At that time, Davy asked Faraday to study and try to understand a discovery that had been made in Denmark and was causing enormous repercussions throughout Europe. This was the experiment by Oersted (1757–1851). The phenomena of electricity and magnetism were known but completely unrelated. Then a lecturer in Copenhagen discovered that when passing an electric current through a wire it deflected the needle of a compass placed over it.[21]

Nobody had been able to explain how this had happened. According to some biographers of Faraday, unusually, his want of a better education curiously favoured him. His lack of formal studies in mathematics and physics prevented him from understanding the complexity of the physics problem being studied. Under these conditions Faraday began to study the relationships between electricity and magnetism in the summer of 1821. In the laboratory he fixed a magnet and imagined various invisible circular lines passing around it. If this were true, a

[21] Maxwell, in his introduction to Faraday's, book reference (Faraday 2003), describes the discovery of the phenomenon of the induction of electric currents by magnetic fields. "In December of 1824, Faraday had tried to obtain an electric current by means of a magnet. On three occasions, he made three fruitless complex attempts to produce a current in a wire by means of a current in another wire, or by means of a magnet. Even so, he persevered. On 12th August 1831 he obtained the first proof that an electric current could induce another current in a different circuit… This was his first successful experiment. During a further nine days of experimenting, he reached the results described in his first series of "Experimental Researches", read before the Royal Society on 24th December 1831".

loosely suspended wire could be dragged around, attracted by the invisible circles. He connected the battery and the wire behaved exactly as he had imagined.

Evidently, there are various hypotheses about Faraday's beliefs, in the sense of trying to explain the reason for his suppositions. Let us overlook these and simply point out that it was in this way that he made the discovery of the century. With the motion of a wire conducting electricity in a magnetic field he had discovered the working principle of the electric motor. As well as this discovery, Faraday managed to link electrical energy with magnetism and, although neither the concept of energy nor its conservation had yet been established, he made an enormous contribution to this by making a connection between electricity and magnetism. Later, Ampère (1775–1836), Gauss (1777–1855) and Ohm (1787–1854) developed theories which would allow them to comprehend profoundly the phenomena of the magnetic fields produced by electric currents and how they flow through conductors.[22]

The discovery of laws which linked the phenomena of electricity and magnetism created favourable conditions on the scientific plane for the Industrial Revolution to reach its second stage. As we know, the first stage of the Industrial Revolution was characterised by use of the steam engine and of coal as a form of energy which replaced water driven machines. The second stage of the Industrial Revolution was the era of electricity and its application in motors and electrical equipment, as well as new forms of transport with the automobile driven by the internal combustion engine. These changes in the means of production had enormous social implications which it is not appropriate to deal with here, even superficially.

In the period we are analysing, that is the end of the eighteenth century until the early decades of the nineteenth century, the physical sciences went through unprecedented transformations in terms of rupture from the Newtonian paradigm, as well opening up promising perspectives with new syntheses and theories explaining physical phenomena linked to light, heat, electricity and magnetism. There follows a synthesis of the most significant events:

(a) Laplace (1749–1827) and his followers had formulated a mathematical theory based on Newtonian mechanics which would be extended to thermal and optic phenomena. However, this theory had to be replaced by new developments in these areas in the decade 1815–25. Despite this, mathematization and the unifying proposal of the physical world contained in the so-called Laplacian project had an important effect on the later development of physical theory.[23]

[22] In September of 1820, in the Academy of Science session following that in which Oersted's experiments were announced in France, André-Marie Ampère published his first observations on the magnetic actions of electric currents. He showed the Academy that electric currents mutually attracted and repelled each other and followed those discovered laws which he called electrodynamics and which were fundamentally important for the elimination of magnetic fluids from science. Georg Simeon Ohm, began his experiments with electric currents in 1825. He used Volta's battery and later replaced it with copper-zinc thermoelectric elements, and could in this way establish the famous law which carries his name. See (Taton 1995), pp. 210~215.

[23] In the next chapter, we approach in more detail, the so-called Laplacian project, with its influences and limitations.

(b) The publication in 1822 of the mathematical theory of heat by Joseph Fourier (1768–1830) brought the study of heat under mathematical analysis, previously applied only to mechanical problems. There was then established a difference between mathematical representation and physical representation. In 1840, influenced by the mathematical analogy between Fourier's heat theory and the theory of electrostatics, William Thompson explained these analogies on one side and on the other the laws of heat and electricity, introducing new views on the mechanics of particles in fluid and an elastic medium.[24]

(c) In 1824, Sadi Carnot wrote a revolutionary work, "Reflections on the Driving Power of Fire", in an industrial context in which the technology of steam engines was of growing interest among French and English engineers. As we shall see in more detail in the chapter dedicated to Lazare Carnot, the model used by Sadi to analyse the principles governing thermal engines owes much to his father, Lazare. Sadi's work ushered in a new science, thermodynamics, initially as a science of machines and later as a general science of transformation processes in nature.[25]

(d) The theory of light waves of Augustin Jean Fresnel (1788–1827), which proposed that light propagates by vibrations of mechanical ether, brought optics into the conceptual framework of the mechanical view of nature. In 1830 this wave theory was accepted and physicists started to explore the great variety of physical and mathematical theories in an attempt to construct a coherent mechanical theory of optics.[26]

(e) In 1827, while examining microscopic grains of pollen suspended in water, the Scottish botanist Robert Brown (1773–1858) noticed that the particles kept moving at random. This phenomenon, observable in both liquids and gases, was given the name Brownian motion. Even though this discovery is

---

[24] Joseph Fourier could be considered the first typical mathematician-physicist. His studies on heat propagation date from 1807, or earlier, and were gathered in a paper presented to the Academy of Science in 1811. His work "*Théorie analytique de la chaleur*", was published in 1822. In the solution to his famous equation on partial second-order derivatives he presents the development of Fourier's series. See (Fourier 1988).

[25] Sadi Carnot was fully aware of the importance of the changes that the development of the steam engine and its employment in diverse branches of the economy would bring for society. In the first pages of his celebrated essay, we read: "The study of these engines is of the greatest interest, their importance is enormous, their use is continually increasing, and they seem destined to produce a great revolution in the civilized world. Already the steam-engine works our mines, impels our ships, excavates our ports and our rivers, forges iron, fashions wood, grinds grain, spins and weaves our cloths, transports the heaviest burdens, etc. It appears that it must some day serve as a universal motor, and be substituted for animal power, water-falls, and air currents. Over the first of these engines it has the advantage of economy; over the other two, the inestimable advantage of power which may be employed at any time and in every place, and will never suffer an interruption in its work". See (Carnot 1990), p. 2.

[26] Frensel's wave theory was a decisive contribution for the abandonment of the imponderable fluids theory. The origin of his work lies in the opposition not only to the scheme of imponderable fluids, but also to Laplace's corpuscular theory of light and the caloric theory for heat. At the start of these works, around 1814, Frensel wrote that he suspected that light and heat were in some way connected with the vibrations of a fluid. His commitment to the concept of light as a kind of motion of a medium, was basic for his optics theory in terms of the motion of wave propagation in a medium, luminiferous ether. See (Harman 1982), p. 21.

not directly related to the rifts we are dealing with, it is important that this fact be registered here, since it is directly linked to another important breach which was to occur more recently with atomic theory.[27]

(f) Although formulation of the law of conservation of energy in 1847 had shaken the unity of physics—if this unity was seen within the above conceptual framework—it had nevertheless, placed the phenomena of heat, light and electromagnetism within the same basis of principles. We shall describe this in detail at the end of this chapter.

A fundamental step in establishment of the conservation of energy principle was the measure of equivalence between mechanical work and heat. In the seventeenth and eighteenth centuries physicists considered energy losses inside mechanical systems as isolated facts arising from non-mechanical processes and so never ventured to formulate a theory of equivalence between heat and mechanical energy. The concept of the conservation of mechanical energy had already been approached by some writers in the eighteenth century. As we saw earlier, Leibniz formulated the principle of the conservation of vis viva. He had affirmed that vis viva, that is, the product of mass times velocity squared, is conserved in the mechanical process.

What must be observed with caution is that the use of the principle of conservation of *vis viva*, despite wide acceptance, did not signify a commitment to Leibniz's theories of nature. This principle was especially applied to the perfectly elastic collision of bodies.

Jean Bernoulli is responsible for the more systematic and consistent accounts on the matter between 1720 and 1730. It is quite true that both he and Leibniz had already perceived that there could be losses in these collisions and they made an analogy between elastic collisions and springs which were prevented from expanding after compression. In this way, they explained the possible losses in energy consumption in the compression of bodies.

Daniel Bernoulli later discussed the operation of a thermal engine which utilised vis viva stored inside coal, by the generation of gases from the coal, but did not suggest any equivalence between heat and mechanical work. Nor did he attribute mechanical losses of the vis viva in inelastic collisions to the heat, as did his predecessors. He continued to consider the losses strictly from the mechanical point of view.

At the beginning of the nineteenth century, the measurement of vis viva through mechanical work, that is, the product of the force times distance, was introduced in various writings of a more technological stamp. This was the case of Carnot. And Peter Ewart (1767–1842) found a quantitative relationship between the heat generated by burning coal and the mechanical power or force which it could produce.

As we shall see in more detail, Lazare Carnot used the conservation of vis viva in mechanical work, although he did not adopt the term work for the product of the force and displacement. It was clear that the work was a measure of energy in

[27] Einstein, in one of five articles published in 1905, would study Brownian motion and define a diffusion coefficient for particles in suspension, on the basis of the dimension of the particles, the viscosity of the fluid, its temperature and the Avogadro number. Following the inverse route, that is, having a way to measure this diffusion coeficient, it is possible to experimentally determine the Avogadro number. See (Les Génies de la Science: Einstein and Maio 2002), pp. 26 and 27.

the field of mechanics, since the sum of these quantities was conserved and if they could be summed up this meant that they were equivalents.

Rumford's experiments (1798) on the heat produced in the machining of cannons showed that a great amount of heat was generated in this process. He then interpreted the phenomenon as arising from motion, that is, from friction, a point of view adopted by Humphry Davy and Ampère. Rumford also obtained a measurement of the mechanical equivalent of heat. This established that one calory was equal to 570 kg-m, with an error of approximately 25 %.

The work of James Prescott Joule (1818–1889) in 1840 was fundamental in clarifying the conversion processes, as well as providing experimental confirmation of this quantitative equivalence between heat and mechanical work. At the start of the 40s of the nineteenth century, Joule's investigations concentrated on the improvement of electric machines and the electrochemical field of investigation already embraced by Faraday as we have seen.[28]

Davy and Faraday had sought to formulate an electrical theory of chemical affinities and Joule also tried to elaborate this theory in the sense of unifying electrical, chemical and thermal phenomena, demonstrating their inter-conversion in the form of quantitative equivalence. In order to arrive at a relationship of this nature between work and heat, he built an electric machine in which mechanical work generated an electric current which, in turn, generated heat; the mechanism provided a numeric relationship between heat and mechanical work to be calculated. Joule soon concluded that mechanical work could be directly transformed into heat by friction.

In 1847, Joule illustrated the direct relationship between mechanical work and heat, in sketches which showed ropes connected to a weight, in such a way that the weight could be raised or lowered according to an increase or decrease in temperature. He did not publish these sketches of motion being transformed into heat, but simply affirmed that the heat was measured through the vis viva and consequently, that particles of the heated bodies were in motion.[29]

If Joule had a fundamental importance in establishing the principle of energy conservation, many others also played a part in this complex process, described in detail

---

[28] Joule's first works concentrated on the perfection of electromagnets, the manufacture of galvanometers and on the properties of voltaic currents. After 1841 he concerned himself with the heat generated by electrical circuits and by electromagnetic machines. It was this work which led to his famous law.

[29] The experimental procedure adopted by Joule consisted in repeating twenty times the process of agitating a liquid by the movement of weights and measuring the final temperature of the agitated liquid. The walls of the recipient containing the liquid were hermetic and made of very thick wood, suitably treated to minimise any heat loss through convection or radiation. His conclusions were as follows: 1) the amount of heat generated by the friction of bodies, whether liquid or solid, is always proportional to the quantity of mechanical work expended; 2) the amount of heat capable of raising the temperature of 1 pound of water (weighed in a vacuum at a temperature between 55° F and 60° F) by 1° F requires for its evolution the expenditure of a mechanical force required by the fall of 772 pounds (350,18 kg) through the space of one foot (30,48 cm). Between 1845 and 1847, Joule repeated these experiments using water, whale oil and mercury, obtaining for these compounds the mechanical equivalents equal to 781.5 lb; 782.1 lb and 786.6 lb, respectively. See (De la Colin 2003), p. 31.

by Thomas Kuhn in his *"Essential Tension"*, first published in English in 1977. Right at the beginning Kuhn affirms: *Between 1842 and 1847, four scientists across Europe, Mayer, Joule, Colding and Helmholtz and, save for the latter, each ignorant of the others work, made public the hypothesis of the conservation of energy. Sadi Carnot, before 1832, Marc Séguin in 1839, Karl Holtzman in 1845 and G. Hirn in 1854, had each written, separately, of their conviction that heat and work are quantitatively equivalent and calculated a conversion coefficient or an equivalent.*

What Kuhn's book proposes to show is a typical case of simultaneous discovery in science and its antecedents, the above citation providing the introductory elements. In addition, he also proposes to explain why, between 1830 and 1850, *so many experiments and concepts necessary for the complete formulation of energy conservation arose to the surface of scientific consciousness.* The analysis which Kuhn makes in his famous article, enumerates three factors which most influenced this simultaneous discovery, namely:

(a) availability of conversion processes;
(b) interest in machines;
(c) philosophy of nature.

The first factor concerns the development of science, the second the process of the Industrial Revolution and finally the third, the ideas and theories of physics at the time. Let us now look at these factors in more detail.

The availability of the conversion processes occurred due to a set of discoveries which began with the invention of the electric battery in 1880 by Alessandro Volta (1745–1827). According to the theory of Luigi Galvani (1737–1798), which was then dominant in France and England, an electrical current could be obtained from the forces expended by chemical affinity, and this conversion proved to be just one link in a much more complex chain which would in time become clearer. Also, as we have already mentioned, Faraday's work, the Oersted experiment and the magnetic properties which could produce motion—something known since antiquity—as well as the fact that it was possible to produce electricity by friction, were all systematically coming together and, little by little, the diverse connections between these phenomena were becoming known.

Furthermore, new experiments and phenomena were yet to be revealed. In 1822, Thomas Johann Seeback (1770–1831) demonstrated that heat applied to a bimetallic belt produced directly an electric current. Twelve years later, Peltier managed to show the inverse process, namely, that an electric current could absorb heat and in this way produce cold. Within this more complex framework showed by Kuhn, Faraday's discoveries were only one more expansion of the field of phenomena in which conversion processes occurred.

Kuhn also called attention to a fact we consider very important for this undertaking, which is the role played by the principle of conservation of *vis viva* for the establishment of the principle of the conservation of energy. He affirms:

> In most histories or pre-histories on the conservation of energy, it is supposed that the mode used to quantify the conservation processes was the dynamic theorem known since early XIXth century as the conservation of "vis viva". This theorem played an outstanding

> role in the history of dynamics, and proved to be a special case of conservation of energy. For that reason, it could well have become a model. Nevertheless, I believe that the prevailing impression that this was the case is erroneous. Conservation of the vis viva was important for Helmholtz to deduce the conservation of energy and a special case—that of free fall—from the same dynamic theorem which so helped Mayer. However, these personages also extracted important elements from another distinct tradition—that of water, wind and steam engineering—and this tradition is all important for the work of the other five precursors who attained a quantitative version of the conservation of energy (Kuhn TS 1996 Chap. 1).

With this citation, Kuhn reinforces one of the viewpoints we have adopted in this study, which is the importance of the expansion of machinery in society and its effects, one of the most notable being the development of general theories for the study of machines, it is not by chance that the second factor considered by him is the *interest in machines*. Kuhn also advances some commentaries about Lazare Carnot:

> Until 1782, in "Essai sur les Machines en General", by Lazare Carnot, the product of the force times the distance did not receive a special name nor a conceptual priority within the dynamic theory. This new dynamic conception of the work concept was not really developed or widespread before 1819–39, when it was fully expressed in the works of Navier, Coriolis, Poncelet and others. All these works are concerned with the analysis of machines in motion. Consequently, work—the whole of force with respect to the distance (product of a force and a distance)—is its fundamental conceptual parameter. Among other significant characteristic results of this reformulation were: the introduction of the term "work" and of the units to measure it; the redefinition of "vis viva" as 1/2 mv2 or mv2/2 to conserve the conceptual priority of the measurement of work; and the explicit formulation of the law of conservation of equality of the work performed and the kinetic energy produced. Only when it was so reformulated, would the conservation of "vis viva" constitute a convenient conceptual model to quantify conversion processes, although almost none of the pioneers used it (Kuhn TS 1996 Chap. 1).

We shall not comment on this long citation from Kuhn, since Chapters five and six are dedicated exactly to the importance of the work concept both in Carnot's general theory of machines and in its development made by personages also mentioned by Kuhn in the citation.

Still following Kuhn's article, we see in general lines the ideas which guided the philosophy of nature of those who discovered the conservation of energy principle. The philosophical current denominated *Naturphilosophie* is of central importance and prime influence for many of those scientists who contributed to the establishment of the principle of the conservation of energy. The word *Naturphilosophie* comes from German Romanticism and expresses a concept which has no linguistic equivalent in French or English. Historians prefer to use the original term in order to designate a specific manner of speculation on the cultural plane where it had a rapid fecund development. It first appeared at the end of the eighteenth century and in the 30 and 40s of the nineteenth century when this kind of thinking was quite strong in German universities.

Kuhn describes the influence of *Naturphilosophie* on the discoverers of the principle of the conservation of energy, thus: *Using the organism as the fundamental metaphor of their universal science, the Naturphilosophers constantly sought a single principle which would unify all natural phenomena. Friedrich Wilhelm Joseph Schelling (1775–1854), for example, sustained that magnetic, electrical, chemical and even organic phenomena must be interwoven, forming a great fusion ... [which] would include the whole of nature.*[30]

Schelling insistently sought the processes of conversion and transformation in the science of his time. Many of his followers dominated teaching in German universities during the first third of the nineteenth century, presenting the new conversion phenomena in a way similar to Schelling's.

It is well known that Oersted was a *Naturphilosoph*, since he persistently pursued the idea of a relationship between electricity and magnetism, driven mainly by his philosophical convictions. In short, many of the scientists who worked in the laboratories of process conversion extracted from their experiments a conception of the physical processes. If we wish to establish connections of other personages with *Naturphilosophie*, we could add that Colding was closely linked with Oersted. Liebig studied for two years with Schelling. Adolf Hirn (1815–1890) frequently cited both Lorenz Oken (1779–1851) and Kant. Robert Mayer did not study *Naturphilosophie*, however, he had close friends linked to these studies. The father of Helmholtz, friend of Johann Fichte (1762–1814), younger than him, was an adept of that philosophy and exhorted his son to abandon strict mechanism.

As we see, contrary to what positivists may think, metaphysics is not only useful in the formulation of scientific theories, but it is also perhaps inevitable that ideas of a purely metaphysical origin may influence scientific thinking. There are numerous examples. As well as this case—the establishment of a fundamental principle of physics—we have many others, like the universal law of gravitation.

John Maynard Keynes (1883–1946) made a citation on the occasion of Newton's third centenary commemorations, which later became famous. He said: *Newton was not the first of the age of reason. He was the last of the magicians, the last of the Babylonians and Sumerians, the last great mind which looked out on the visible and intellectual world with the same eyes as those who began to build our intellectual inheritance rather less than 10,000 years ago. Why do I call him a magician? Because he looked on the whole universe and all that is in it as a riddle, as a secret which could be read by applying pure thought to certain evidence, certain mystic clues which God had laid about the world to allow a sort of philosopher's treasure hunt to the esoteric brotherhood.*

This citation is made due to a series of studies and documents showing that Newton's thinking was much more influenced by alchemy than previously imagined. In the thinking on the history of science a positivist view predominated, represented chiefly by George Sarton. This is why those words proffered by Keynes in that faraway year of 1946 still cause so much commotion in scientific circles.

Closing these considerations on the principle of the conservation of energy, we should observe that if the ideas on the unification of natural phenomena coming

---

[30] It is common to consider the influence of Immanuel Kant (1724-1804) and even that of Leibniz and dynamism on Schelling. For the first two thinkers, the first concept is the inherent force of matter.

from *Naturphilosophie,* on the one hand played a fundamental role in this new synthesis in physics, uniting diverse phenomena from different fields into a more general principle, on the other, this discovery also fomented new conceptions and strengthened a dynamistic view of nature. In reality, this was the subjugation of the Newtonian paradigm which was spreading through diverse fields of physics, very often with the use of implements developed by Newton himself. This is the case, for example, of André-Marie Ampère (1775–1836), who affected the first physical–mathematical synthesis between electricity and magnetism in his *Memoire sur la Théorie Mathématique des Phénomènes Électrodinamiques Uniquement Déduite de l'Expérience,* published in 1827.

It should also be pointed out that the subjugation of a paradigm with the range and explicative power of the Newtonian paradigm was not made without some resistance. In 1867, almost two centuries after the first edition of Newton's *Principia*, William Thomson and Guthrie Tait published their *Treatise on Natural Philosophy* with great success and wide repercussions in scientific circles. In it the authors tried to *update* Newtonian mechanics and went on to defend the idea that Newton's third law (law of action and reaction) is a much more general law of conservation, being a forerunner of the principle of the conservation of energy.

References

Actes de La Journée Maupertuis (1975) Creteil 1973. Librairie Philosophique J. Vrin, Paris
Aristotle LP (1999) Librairie Philosophique J. Vrin, Paris
Atkins P (2003) Galileo´s Finger. Oxford University Press, U.K
Carnot S (1990) Réflexions sur la Puissance Motrice du Feu. Éditions Jacques Gabay, Paris
D´Alembert JL (1921) Traité de Dynamique. 2 Vol, Gauthier-Villars et Cie. Éditions, Paris
De la Colin LG (2003) Maquina de Vapor al Cero Absoluto. Fondo de Cultura Economica, México
de Maupertuis PLM (1744) Accord des Différents Lois de La Nature qui Avaient Jusqu´ici Paru Incompatibles, Mémoires de l Academie des Sciences de Paris
Dugas R (1988) A history of mechanics. Dover Publications, Inc., New York
Duhem P (1997) Les Origines de la Statique, Paris, 2: 5–121
Faraday M (2003) A História Química de uma Vela. Editora Contraponto, Rio de Janeiro
Fourier JBP (1789) Memoire sur la Statique Contenant la Démonstration du Principe des Vitesses Virtuelles, Journal de l´École Polytechnique, 5eme cahier, Paris
Fourier JBP (1988) Théorie Analytique de la Chaleur. Éditions Jacques Gabay, Paris
Gueroult M (1967) Leibniz. Dynamique et Métaphysique, Aubier Éditions Montaigne
Harman PM (1982) Energy, Force and Matter: The conceptual development of nineteenth-century physics, Cambridge University Press, UK, USA
Heath TL (2002) The works of archimedes, Dover Publications, Inc., N. York
Huygens C (1973) Horologium oscilatorum, translated from latin by Jean Peyroux, Librairie Albert Blanchard, Paris
Lagrange JL (1989) Mécanique Analytique. Éditions Jacques Gabay, Paris
Les Génies de la Science: Einstein, Maio 2002–Agosto 2002, Edições do CNRS
Lindsay RB (1975) The concept of energy and its early historical development. Brown University, Pennsylvania, p 12
Locqueneux RP, Histoire de la (1996) Thermodynamique classique (Une Histoire de La Chaleur), NO. 45, Decembre 1996, Librairie A. Blanchard, Paris

Moreira IC (1998) Maupertuis e o Princípio da Mínima Ação. Publicação do Instituto de Física da UFRJ, Rio de Janeiro

Oliveira ARE (2004) The contribution of coulomb to applied mechanics, Proceedings HMM 2004 Symposium, Kluwer Academic Publishers, p 217–226

Planck MA (1993) Survey of physical theory, Dover Publication, Inc., N. York

Poinsot L La (1975) Théorie Générale de l'Équilibre et Du Mouvement dês Systèmes, Édition Critique et Commentaires par Patrice Bailhache. Librairie Philosophique J. Vrin, Paris

Séris JP (1987) Machine et Communication: Du Théatre des Machines a la Mécanique Industrielle. Librairie Philosophique J. Vrin, Paris

Taton R La (1995) Science Contemporaine. Quadrige, PUF

Witkowski N (2004) Uma História Sentimental das Ciências. Jorge Zahar Editor, Rio de Janeiro

Chapter 4
The Rational Mechanics Dilemma

> *I wish we could derive the rest of the phenomena of nature by the same kind of reasoning from mechanical principles, for I am induced by many reasons to suspect that they may all depend upon certain forces by which the particles of bodies, by some causes hitherto unknown, are either mutually impelled towards one another, and cohere in regular figures, or are repelled and recede from one another.*
>
> (Isaac Newton—Preface to the first edition of *Principia, May* 8, 1686)

4.1 A Brief History of Mechanics in Physics Context

Since its genesis, the study of mechanics has always been associated with geometry and oftentimes with astronomy. It is unnecessary to say that any kind of mechanical theory is based on some concept of space. Specifically, classical mechanics and the special theory of relativity use Euclidian space, which is the place where motion occurs. Its metric is given by the distance between two points calculated by the Pythagorean theorem. With respect to astronomy, which is at the center of the seventeenth century Scientific Revolution, we can say that, historically, it gave origin to mathematical physics, as remarked by Poincaré: *Mathematical physics, as we know, was born from celestial mechanics at the end of eighteenth century, when it achieved its complete development. In his first years, although the child looked like his mother in a very impressive way* (Poincaré 1970).

Let us follow the long history of mechanics in a concise way. The pioneering mechanical studies were made by Aristotle and Archimedes, as we saw in the previous chapter, in spite of certain old documents of unknown origin. The two thinkers lived in related epochs but in different contexts. Aristotle lived in the period of

A. R. E. Oliveira, *A History of the Work Concept*, History of Mechanism and Machine Science 24, DOI: 10.1007/978-94-007-7705-7_4,

Greek history that led to the end of state-cities and Archimedes lived in the Alexandrian period.[1]

The fundamental differences between Aristotle and Archimedes regarding the development of mechanics are that Aristotelian mechanics are part of his philosophy. In other words, when Aristotle tries to characterize motion, it is done within the framework of general changes and transformations. Aristotelian mechanics are completely qualitative and developed in a historical and speculative context of philosophical thought of which he was one of the most important adherents. In a world dominated by social hierarchies, mechanical motion would be too. To him, motion was classified into two types: the *natural* and the *violent*. In the first, the body tends to go back to its natural position (to heavy bodies, the center of earth); the second type occurs when the body is displaced from that position (for instance, the motion of projectiles).

Archimedian mechanics, as well as the whole of mechanical studies of the Alexandrian period, are characterized by extreme specialization, attacking very specific problems within a new historical context created by the expansion of Alexander's Empire. One should emphasize the deep economic reforms, the changes in currency, the trade development and the extended exchange among countries throughout the Mediterranean Sea. This new context also created new social demands, including military activities. It was not by chance that Archimedes developed mechanical principles, such as the principle that bears his name, as well as several types of military devices. Other inventions and stories are attributed to him without proof or confirmation.[2] And yet, in the Alexandrian period, we have more two important figures: Pappus and Hero of Alexandria.[3]

After the decline of the Alexandria School, the development of mechanics would pass through a long period of lethargy. At the same time, a great expansion would occur within the Arab world. During this period of Arabian domination,

---

[1] Bertrand Russel (Russel 1967), vol. 1, p. 249 states: *The trajectory of Greek language world, In ancient times, could be divided in three periods: free State-Cities and those finishing with Felipe and Alexander; the Macedonian domain and its last vestige with roman annexation of Egypt, after Cleopatra's death and the last, the Roman empire. The first one is characterized by freedom and disorder, the second by submission and disorder and the third by submission and order.*

[2] And yet, the same author in the same volume states on page one: *The second of these periods is known as Hellenic age. In sciences and mathematics, the work accomplished during this period is the best already by Greeks. In philosophy, it includes the foundation of epicurean schools and stoics, as well as the skepticism as a doctrine definitively formalized; it is, although, still philosophically important, but less than in the Plato and Aristotle period.*

[3] We have already referred to Hero in the previous chapter. Pappus lived in the fourth century and, according to René Dugas (Eves 1997), seems to be the one ancient mathematician to study equilibrium and the motion of a body along an inclined plane. These studies inspired scientists from the Renaissance to continue the work of Guido Ubaldi and Galileo. A study about the center of gravity, by tracing lines with a suspended body using several points of suspension, is also attributed to Pappus.

mechanics showed low progress with other fields of physics, such as optics, being developed instead.

This situation began to change in the thirteenth century. It is now easy to see the flourishment of philosophical, theological and scientific studies that took place at this time. Not only were some of these studies developed by Arabs but they also translated a great quantity of scientific and philosophical texts from Greek. It is important to note in this context the appearance of Universities with characteristics quite different from those of traditional institutions.[4]

In the fourteenth century, a new and fundamental change would occur in the established conception of motion, namely, the disappearance of the spirits from the physical world. This was an important step in the understanding of the modern law of inertia. One of the first to think in this direction was Jean Buridan (1300–1360) in the middle of the fourteenth century. A few years later, his disciple Nicole Oresme (1350–1382) achieved a much better understanding than that of Buridan. He stated that one could suppose that God had put the universe to work as one would a kind of watch and that since then it has been left to work by itself. Metaphysics, step by step, allowed for physical phenomena to develop by itself, which is important to the understanding of the laws of motion.

Thus, Buridan achieved an important concept: the doctrine of *impetus*. It is an important advancement because it is the beginning of the solution of the contradiction within Aristotelian physics, which appear in the study of projectile motion. In other words, if we look at Aristotle's statement that motion implies a driven force besides the resistance imposed on motion, in fact, *impetus* replaces the force eliminating this enormous difficulty. In addition to the contribution of Buridan, Oresme has developed an interesting diagrammatic form to represent a variation of a given quantity, and this representation was later applied to a graphic visualization of motion. This fact is also considered an important step in the construction of a system of coordinates.[5]

---

[4] In (Corbisier 1988), p. 153, one reads: *The word Universitas did't designates, in middle ages, some settled schools in the same city, but the community of students and lecturers that participated of teaching in the same place. The "studium generale" or "universale", was a center to which converged students from any remote country´s city. Such study centers were founded by initiative and sponsoring of church, in the cities considered important from religious point of view.*

[5] In (Dugas 1988) in Chap. 3, p. 59, one reads: *We will follow the Tractatus de figutatione potenciarum et mensurarum differentiarum. Oresme starts from the principle that everything that we can measure can be thought as a continuous quantity. Each intensity can be represented by a straight line plotted vertically from each point of the "subject" that changes the intensity.* In other words, the quantity is represented on the horizontal line or extension which affects the quantity the values of which were represented vertically. Using this approach, Oresme built kinematic diagrams, representing time as an extension on the horizontal line and dividing it into proportional parts, so that a geometrical progression of ratio ½ is built, where the first term is *t/2*. Hence, it is possible to calculate the velocity in any motion interval and the total distance performed.

We should mention the contributions to mechanics that came out of the Oxford School in the same period. Based on their studies in logic, some important results were obtained. One of the most important figures was William Heytesbury, who, around 1371, was chancellor of Oxford University. His greatest contribution was the concept of acceleration, an unknown concept to Buridan and Oresme at the Paris School.[6]

At the beginning of the fourteenth century, an important social and cultural change arose in Italy: the beginning of the Italian Renaissance. One characteristic of this movement was its humanism. By humanism, we mean the importance of classical studies, as well as the consideration of classical antiquity as the model and paradigm for all cultural activities. A figure that epitomized this period was Francesco Petrarca (1304–1374).

In the fifteenth century the Italian school of mechanics emerged with important contributions. The most important figures were: Blasius de Parma (1327–1416), Nicholas of Cues (1404–1464) and Leonardo da Vinci (1452–1519). Blasius wrote a *Treatise on Weights* and *Quaestiones super tractatu de latitudinibus formarum*, the latter printed in 1486. He was also responsible for the introduction into the Italian school mentioned above of the statics of the thirteenth century and the kinematics of the fourteenth century. Nicholas de Cues was originally a metaphysician. He became bishop of Brixen in 1450 and his works were published in three parts between 1500 and 1514. On the subject of mechanics, he left a number of dialogues entitled *De ludo globi*. There is an approximation between the ideas of Nicholas de Cues and the doctrine of *impetus* of the Parisian Schoolmen. To him, God should have imparted an *impetus* to the spheres (his model) at the beginning and that *impetus* would then be conserved indefinitely.

Leonardo da Vinci is one of the most interesting figures in the history of mechanics. He lived in the period immediately before the Scientific Revolution. He is an engineer, artist, scientist and inventor. He studied many different problems, including: the application of the concept of the moment of a force, rigid body motion on an inclined plane, solution of a system of forces, energy of a body in motion, study of the earth's shape, the theory of the center of gravity, the falling of bodies, hydrostatic and hydrodynamic problems and many others (Fig. 4.1).[7]

In the seventeenth century, very close to the Scientific Revolution, a very interesting figure appeared in the history of mechanics: Dominic de Soto (1494–1560), of Spanish heritage. His importance springs from the fact that his fundamental

---

[6] In Ref. (Dugas 1988) in Chap. 3, pp. 66 and 67, one reads: *Heytesbury was professor of Merton College in 1330, belonging to Queen's College around 1340, being also Chancellor of Oxford University in 1371...In his treatise De Tribus Praedicamentis, Heytesbury has distinguished between "latitude motus" (velocity) and the "velocitas intensionis vel remissionis motus" whose value was the positive or negative increment of the last. This quantity corresponds to acceleration.* As we will see later in this chapter, who introduces the concept of acceleration in France is Pierre Varignon.

[7] Leonardo da Vinci, Jerome Cardan (1501–1576) and Simon Stevin studied the possibility of a machine with perpetual motion, and they concluded that it is impossible. In this sense, they made an important contribution to the discussion of the energy conservation problem.

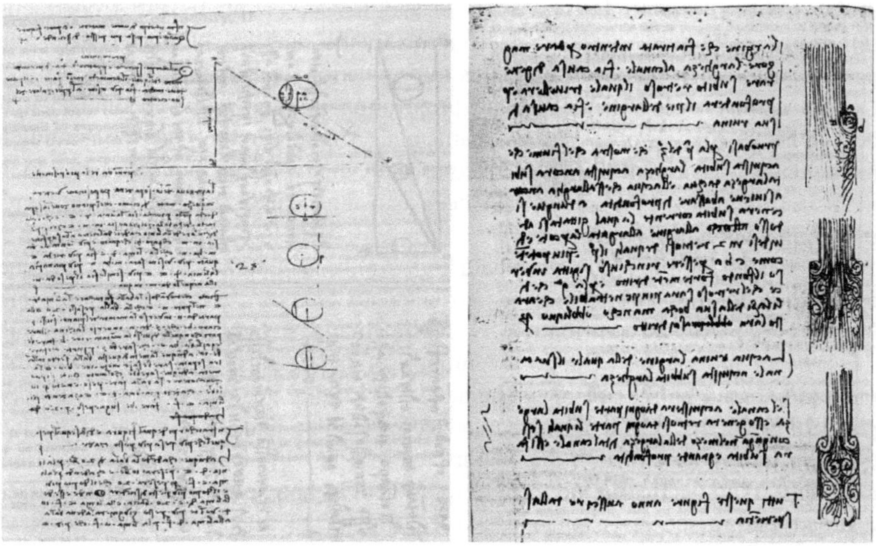

Fig. 4.1 Leonardo da Vinci sketches

work deals with the falling of bodies as we know it from one of the problems solved by Galileo by the mathematization of space. Others figures to appear in this period in Italy were Nicholas Tartaglia (1500–1557) and Bernardino Baldi (1553–1617). Tartaglia conducted studies on dynamics, mainly the motion of projectiles, and Baldi studied equilibrium problems as the stability of balance.

Two other prominent figures to appear in Italy were Guido Ubaldi (1545–1606) and Giovanni Benedetti (1530–1590). Ubaldi was, in his time, a great authority on mechanics, being considered one of Galileo's teachers. His name appears many times in Lagrange's *Analytical Mechanics*. He also systematically criticized the Schoolmen of the thirteenth century. Benedetti studied practically the entirety of mechanics of his time and published an extended work in 1585 entitled *Diversarum speculationum mathematicarum et physicarum.* In this work, circular motion is treated. He wrote that if a body is released from circular motion, it will travel in a straight line which is tangent to the original circle of motion. This result also anticipates some of the achievements of Huygens (1629–1695) and Newton. Also important is the way Benedetti's ideas anticipate those of Galileo. He stated that bodies composed of the same material fall at the same speed regardless of their weight. He justified his claim with an argument using Archimedes' results on bodies in a fluid.

Further developments in mechanics are related to the seventeenth century Scientific Revolution. In spite of its significant social and cultural changes, for our purposes, we will only briefly summarize its scientific transformations. It is possible to distinguish the first step of the Scientific Revolution as being a significant change of perspective in the cosmological vision associated with Copernicus

(1472–1543), Ticho Brahe (1546–1601) and Kepler (1571–1631).[8] Copernicus reformulated Ptolomy's (100–170) solution to the problem of planets with the need to restore their lost harmony. This reformulation was accepted by Kepler and Galileo mainly owing to its realistic proposition. Kepler, with the fundamental help of Ticho Brahe's observations, mathematizes nature in the sense of creating a mathematical physics of the heavens, completely distinct from the model on which many astronomers' best efforts had been expended since the time of Ptolemy, including Copernicus himself.

The second step accepted for characterizing the Scientific Revolution was the establishment of scientific method attributed to Galileo's work on motion and Newton's subsequent contribution by systematizing, generalizing and extending the laws of motion on earth to the motion of the solar system.[9] The scientific method is based on the use of mathematics, especially geometry, for the study of physical problems. Nature as mathematized by Galileo finds itself represented at a different level of abstraction than nature realized in daily experience. This difference appears to Galileo in the form of a problem of how to ensure that the mathematically expressed laws he found were valid in the same way at the level of experience as well. The necessary means for establishing this were discovered by Galileo in experiments. He had already used "mental experiments" to heuristically explain mathematical regularities that are behind free fall and projectile motion. Experiments also appeared to Galileo to provide a means by which to bridge the gap between phenomena represented at an idealized and at an empirical level (Fig. 4.2).

---

[8] In (Kuhn 1957), p. 20, one reads: The astronomical reform, is not, although, the unique finality of the publication of Copenicus's *De Revolutionibus*, in 1543. Some of these innovations, which prepared one century and half later Newton's conceptions of universe, were the unforeseen results of Copernicus's astronomical theory. Copernicus has suggested earth motion with the purpose to improve the techniques used to predict astronomical positions of celestial bodies. To other sciences, his suggestion has created new problems and until to be solved the astronomer's conceptions about the universe were not compatible with the science men. During the seventeenth century, the reconciliation with Copernicus's astronomy was an important fact to intellectual fermentation of the Scientific Revolution. In this development, science has performed a new role, and since that, it is a prominent factor to society development and to western thought.

[9] (Ludovico Geymonat 1997), analyzing the revolution of scientific method provided by Galileo, states: *The strong appeal to experience did not consists certainly in something radically new in Galileo time, or a peculiarity of his methodology* ...(p. 322*) It is universally recognized that the main innovation introduced by Galileo into scientific methodology, consists in the broad use of mathematics as indispensable tool to nature knowledgement* ... (p. 311) *The point which we need emphasize, is other: is his clear intuition about the fundamental importance to scientific observation of phenomena is to perform measurement. This, and only this, in fact could describe with precision the phenomena by differentiating the experiences which seem identical to naïve observer. It permits, in addition—when observed data are translated in numbers—to classify experience results in a mathematical discourse not only very rich, but strongly rigorous than common discourse. We have, then, no more a qualitative physics, but a quantitative one, whose superiority when compared with the old traditional science became dominant in any field.* (p. 329).

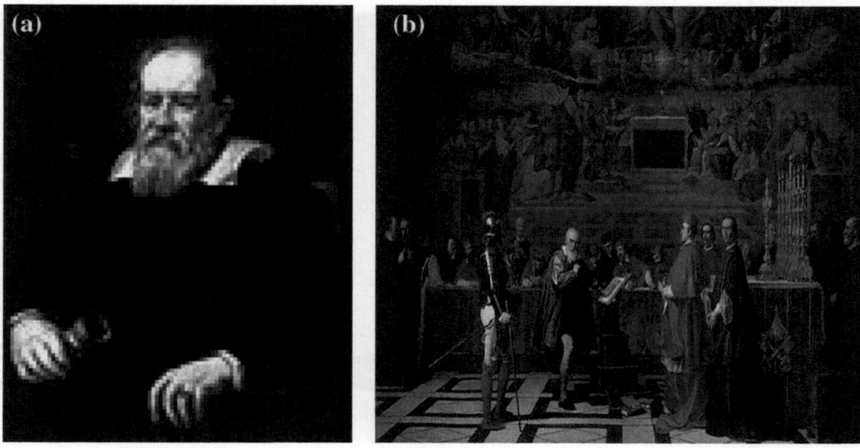

Fig. 4.2 **a** Galileo Galilei (1564–1642). **b** Galileo's trial

A few words about Newton's *Principia*. No other work in the history of science has equalled this book, either in originality, power of thought and investigation capacity or even in terms of achievement. Indeed, no other work has similarly transformed the structure of science; consequently *Principia* has no precursor. Newton's purpose was to insist that the range of events in nature may seem to be almost infinite in appearance but that, in reality, all phenomena and their properties must be traceable to a small set of fundamental laws of nature and by mathematical reasoning each of them is deducible again from these laws (Fig. 4.3).

Yet, with respect to the Scientific Revolution, we should mention the name of Descartes mainly in association with the fight against scholastics and through his offering of an alternative vision to the world known as mechanicism. Also, the new perspective that he adopted using doubt systematically as a method mortally wounded the vital principles of scholasticism: the dogma and the authority principle. Descartes was also important in the building of basic concepts of mechanics, as we have previously seen, when he postulated the principle of the conservation of quantity of motion in the universe instead of *vis viva* as proposed by Leibniz. His contribution to Newton's laws of motion was essential. It is remarkable that the concept of force that dominated dynamics in the seventeenth century before the work of Newton was a legacy from *impetus* mechanics and the common sense perceptions which it formalized. It was a legacy ultimately beyond reconciliation with the principle of inertia that was basic to the new science of motion. The idea of inertia appears to have demanded a concept of force measuring not motion but change of motion such as Newton proposed. Newton's statement of inertia is derived immediately from Descartes. He had drawn from the principle of inertia the conclusion of the relativity of motion, a conclusion impossible to reconcile with the universal dynamics toward which Newton was groping.

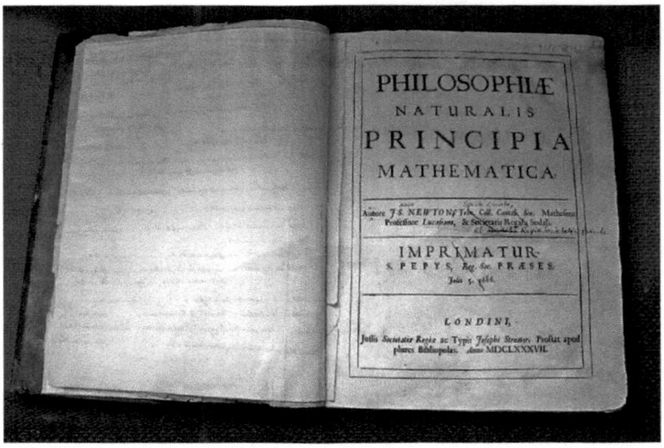

Fig. 4.3 Newton's own Principia book

The appearance of no classical mechanics began with the attempt to apply the principle of Galilean relativity to the other fields of physics. We use, as an example, electrodynamics (Ronan 1987). The experiments of Albert Michelson (1852–1931) and Edward Morley (1838–1923) had shown that light velocity was the same regardless of the relative motion of the observer as well as the source in all directions. Thus, the Galilean transformation cannot be correct and must be replaced by another transformation which maintains the invariability of light velocity in all systems. Such transformation is known as a Lorenz transformation, which can be applied to physical phenomena, mechanical or electromagnetic. With these results, Albert Einstein, in 1905, postulated a series of statements, coherent with the constancy of light velocity. These changes contain the basis of the special theory of relativity (Einstein 1999). The most important of them is the rejection of absolute time, and since that, time has been considered a fourth coordinate in tridimensional Euclidian space.

The development of the theory of relativity occurred with the attempt to eliminate inertial forces which appear when the observer is associated with a rotating system of reference. These forces are kinematic in nature and can be eliminated using the Galilean system of reference. However, the gravitational forces cannot be eliminated through these transformations, and a new theory arises. This new theory is called the gravitational theory of Einstein or the general theory of relativity (La Mason 1986).

The rise of quantum mechanics is directly associated with the solution of the problem known as ultraviolet catastrophe, also called the Rayleigh-Jeans catastrophe. Max Planck (1858–1947) solved this problem postulating that electromagnetic energy did not follow the classical description, but could only be emitted in discrete packets of energy proportional to the frequency, thus formulating Planck's law of body radiation. Based on past experiments, Planck was also able to determine the value of its parameter, now called Planck's constant. The packets of energy later came to be called photons, and played a key role in the quantum description of electromagnetism (Brenan 1998).

With this brief description, we summarize the history of the theories of mechanical phenomena in the context of physics in general, but also associated with scientific and philosophical ideas.

4.2 Formalization of Mechanics in the Eighteenth Century

During the Eighteenth century, Newtonian mechanics was significantly changed with respect to its formalization. With this process undertaken by Lagrange, we could say that mechanics was not only renewed but enriched and transformed, acquiring the denomination of classical mechanics. This will be described in the following sections.

4.2.1 Background

The formalization of mechanics by Lagrange was due to the development of differential and integral calculus through the pioneering works of Newton and Leibniz, as is well-known, but the contributions of many others mathematicians were integral as well. It is important to establish certain connections between the development mentioned above and other advances in correlated areas in order to have a better understanding of the progress of mechanics within the wider process described briefly in the last section. If we look at Lagrange's work for achieving a new formalization of Newtonian mechanics, it is easy to see that he wished to perform a double reduction in motion: from mechanics to analysis and from analysis to algebra.[10]

With respect to the history of calculus, the first to arise was integral calculus and then, after a long period of time, differential calculus. The first appears in several processes of adding figures, in order to obtain lengths, areas and volumes. The differentiation was created more recently to solve problems involving tangents of curves, as well as in the search for the maximization or minimization of functions. After both developments, it was proved that integration and differentiation are interrelated such that one is the inverse operation of the other. Despite having happened in the seventeenth century, this achievement is also related to Greek mathematics. The names of Eudoxus and Archimedes[11] are associated with the history

[10] See chapter II of (Dalmenico 1992), p. 29, entitled *The analytical ideal of Lagrange*.

[11] In (Dalmenico and Pfeiffer 1986), p. 171, one reads: *Eudoxus's theory of proportions (born around 408 B. C.), presented in book V of Euclid's Elements is an attempt to give validity to incommensurable quantities and witnesses in some way the introduction of irrational numbers into the field of Greek mathematics. It is in the basis of exhaustion method, which will permit to Greeks to solve problems, only solved later, belonging to infinitesimal calculus: calculation of curves lengths, calculation of areas or volumes determined by curves or curve surfaces, determination of gravity centers, tangents determination,* etc.

of calculus due to certain methods and techniques that they developed in order to subdivide figures, being the reasoning behind these methods under infinitesimal operations or the approximate calculation of areas and volumes of figures. These operations also contain the idea of passage to limit.

From later times, Bonaventura Cavalieri (1598–1647), a disciple of Galileo's, can be considered one of the first precursors of calculus. In his work *Geometria indivisibilibus*, published in 1635, the *indivisible method* is presented. This is a very long treatise that does not exactly put forth the concept of the *indivisible* clearly. One can interpret the indivisible of an area as a chord of this area while the indivisible of a solid would be a section of it. The achievements obtained in these studies have led to Cavalieri's known principles. With this contribution, an important step towards the subdivision of plane or solid figures had been taken, as well we a fundamental step towards the integration method.

The differentiation process was originated by methods for the determination of the maximum and minimum of functions, tracing tangents to these curves. In spite of certain Greek mathematicians having attacked similar problems, a clear operation that we can consider a differential method appears with Pierre Fermat (1601–1665), who expressed these ideas in 1629.[12]

Kepler also remarked that increments of a function became infinitesimals in the neighborhood of a common point of the maximum or minimum. Fermat used this fact as a method to determine these points (Eves 1997).

Before the establishment of Newton and Leibniz's calculus, two immediate precursors of Newton were: John Wallis (1616–1703) and Isaac Barrow (1630–1677). Wallis was the first to discuss conic curves by means of an algebraic second degree equation instead of considering a conic's section. In 1655, he published *Arithmetica infinitorum.* In this book, the methods presented by Descartes and Cavalieri are systematized and extended. Barrow was born in London in 1630, completed his studies in Cambridge and was an expert in Greek culture. He was succeeded by Newton in his chair by his own wish. Barrow's masterpiece is *Lectiones opticae et geometricae*, which appeared in 1669/1970. In this book, we find an approach very similar to the modern differential process through the use of the *differential triangle*, very common in calculus textbooks. And yet, the discovery that differentiation and integration are inverse operations is attributed to Barrow. This is known as the *fundamental theorem* of calculus, appearing in the above-mentioned *Lectiones*.

4.2.2 The Tools

After the mathematical achievements of the seventeenth century, the infinitesimal methods had growth and thus arose the possibility of systematizing and ordering

---

[12] In (Eves 1997), p. 429, one reads: *In spite logical process proposed by Fermat be incomplete, it is possible to identify that his method is equivalent to* $\lim f(x+h) - f(h) = 0$.

them. This task was done by Newton and Leibniz, the latter, a Jesuit, philosopher and German politician. Through independent means, they invented the algorithm procedure and established the connections between several problems that appeared to be separate. Hence, they built an autonomous branch of mathematics and are considered the founders of differential and integral calculus.

Newton wrote three different works on calculus. They were published at the beginning of the eighteenth century and had a restricted influence because Newton hesitated throughout his life to publish his findings. A first mention of his fluxions theory appears in 1687 in *Principia.* The propositions about speed, acceleration, the tangents of curves, all these definitions in geometrical forms were a strong encouragement for research into infinitesimal calculus.

Newton had three different conceptions about infinitesimal calculus. (1) The first was the infinitesimal conception, with the direct influence of Barrow and Wallis. He operated with infinitesimal quantities called *moments*, which are equivalent to Fermat's increments; (2) The second was the method of fluxions, the most famous, where he considered the mathematical quantities as generated by continuous increments, in a similar way to the space described by a body in motion. Hence, the variations of those quantities are equivalent to the speeds of generated motions. These speeds are called *fluxions* (Newton 1994) in Chap. 2; (3) The third and last conception is known as the method of first and last reasons. It appears in his *Quadratura curvarum*, written in 1676 and published in 1704. Newton tried to eliminate any clue of infinitesimal quantity, by initially considering only its relationships.

Despite their differences, Leibniz's contribution to the establishment of infinitesimal calculus is of no less importance than that of Newton. After studying advocacy and philosophy, Leibniz began participating in German politics, going on a diplomatic mission to the court of Louis XIV. During his time in Paris, he knew Huygens (1629–1695), a member of the Academy of Sciences, which had only recently been created. Under the influence of Huygens, he initiated the study of Cavalieri, Roberval, Pascal, Descartes, Gregory and Wallis.

In 1676, Leibniz left Paris to follow his diplomatic carrier in Hanover. There, he published a series of short articles about differential calculus which first appeared in 1684 in *Acta Eruditorum*, a scientific journal founded with his support in the same year in Leipzig. Unfortunately, some of these works have never been published.

The advantage of the Leibnizian method in comparison to the Newtonian is the simplicity of his algorithm, the elegance of his notation, and the best operatory manipulation permitting the calculations to be performed in a practically automatic way independent of the nature of the objects under consideration. In addition, his method contains formal elements associated with geometrical analogies. He also eliminated the metaphysics from the infinitesimal analysis by considering them as aiding elements in the same manner as imaginary numbers are used in mathematical analysis. In the absence of rigorous definitions, Leibniz interpreted the infinitesimals in terms of instantaneous variations, as did Newton.[13]

[13] For the majority of Leibniz's articles regarding infinitesimal calculus, (see Leibniz 1983). *Analyse des infinement petits, pour l´intelligence des lignes courbes* (1696).

In spite of these achievements, the generality of methods, and the algebrization of the calculations, the development of infinitesimal calculus was not yet based on solid ground from the conceptual point of view. Fundamental concepts, such as limit, differentiation and integration, were not well defined, although this was merely a question of more time. Some other contributions would be added to the founders' work. The first to use as well as develop this new mathematical tool was Jacques Bernoulli (1654–1705). After his nomination in 1687 to Basel University, he contacted Leibniz by letter, addressed December, 15th 1687, asking him to clarify certain aspects of the new calculus. Because of his travels in Germany, Austria and Italy, Leibniz did not answer him until three years later on December, 24th, 1690. In the interim, Bernoulli was not wasting time. He studied and assimilated calculus, leading to May 1690, when, in *Acta Eruditorum*, he applied calculus for the first time, solving the problem proposed by Leibniz in 1687: the isochronous curve, in which the vertical component of the velocity of a particle rolling over this curve is constant. At the same time, Jean Bernoulli, with the help of his brother Jacques, also began to study the new calculus. Starting in 1690, the two brothers Bernoulli worked together with the help of Leibniz, to apply the new methods to different problems successfully.

A curious development then occurred. In a period in Paris during the winter of 1691/1692, Jean Bernoulli began to instruct the Marquis of L'Hopital in Leibniz's calculus. With this interchange, the Marquis wrote the first treatise on differential calculus, published in Paris at the end of June 1696, entitled *Analyse des infinement petits pour l'intelligence des lignes courbes*. The author, in a letter addressed to Jean Bernoulli, written on July 15th, 1696, left to others the task of developing and to writing about integral calculus (Fig. 4.4).

(a) (b)

Fig. 4.4 **a** Marquis of l'Hôpital. **b** First book of differential calculus

Marquis of l'Hopital is one of the most important figures of the mathematicians of the Malebranche (1638–1715) group. Reyneau, Jacquemet, Bernard Lamy, Varignon and Fontenelle also belong to this group. The works of Andre Robinet and Pierre Costabel, quoted by Michel Blay (Blay 1992) in Chap. 2, have attributed a kind of reform of the Cartesian mathematics to this group. It consisted of the assimilation of new Leibnizian methods for the introduction of infinitesimal calculus into France.

Michel Blay also notes the pioneering work of Pierre Varignon, who, starting from Leibnitzian algorithms, built the concepts of instantaneous velocity and acceleratory force. These new results appeared in two memoirs for the Royal Academy. The first on July 5th and the second on September 6th, both in 1698:

(a) *Regles generals pour toutes sorte de mouvements de vitesses quelconques variées à discrétion.*
(b) *Application de la règle générale des vitesses varies, comme on voudra, aux mouvements par toutes sortes de courbes, tant mécaniques que géometriques, d'où l'on déduit encore une nouvelle manière de démontrer les chuttes isochrones dans la cycloide renversée.*[14]

4.2.3 Lagrangian Mechanics

As previously noted, Newtonian mechanics underwent a profound change through the formalization carried out by Lagrange. Through the work of Lagrange, rational mechanics reached the position desired by Cartesians of becoming a branch of pure mathematics. Some years before, at the beginning of the eighteenth century, differential and integral calculus had been sufficiently developed, becoming a useful tool for the solution of a series of problems in physics and mathematics. Hence, the conditions for the application of the new analytical and algorithmic procedures to the science of motion were given.

In 1736, Euler wrote the first treatise on mechanics of the material point,[15] known as *Analytice exposita*. Some years later, in 1743, d'Alembert explained his mechanical philosophy in a preliminary discourse to his famous *Traité de Dynamique*. D'Alembert had a different conception of force, as a derived notion, in complete disagreement with Newton. In addition, he attributed the fundamental importance to the concept of mass as well as all pure kinematic elements.

---

[14] The first memoir, according to (Truesdell 1983) in Chap. 2, had the objective: *To provide a general expression to the velocity, which permitted the treatment to any motion with straight trajectories, for any mode of velocity variation.* Yet according to the same author: *The construction of the concept of instantaneous velocity is a fundamental step. Although the culmination of this process is the establishment of the concept of instantaneous acceleratrix force; then, the expressions of these concepts can be derived one from the other by simple calculation using leibnizians algorithms.*

[15] The complete title of Euler's work is: *Mechanica sive motus scientia analytice exposita.* In fact, it is a kind of investigation program. Euler, after reading the work of the founders of mechanics, mainly Huygens and Newton, adopted the purpose of transforming mechanics into a rational science starting from the definitions and trying to organize its propositions.

MÉCHANIQUE
ANALITIQUE;
Par M. DE LA GRANGE, de l'Académie des Sciences de Paris, de celles de Berlin, de Pétersbourg, de Turin, &c.

A PARIS,
Chez LA VEUVE DESAINT, Libraire, rue du Foin S. Jacques.

M. DCC. LXXXVIII.
AVEC APPROBATION ET PRIVILEGE DU ROI.

Fig. 4.5 **a** Joseph-Louis Lagrange (1736–1813). **b** Analytical mechanics

The reading by Lagrange of Euler's treatise "Methodus inveniendi lineas curvas maximi minimine proprietate gaudentes", published in 1744, led Lagrange in 1755 to discover his "method of variations", which Euler (1707–1783) had sought in vain. Lagrange communicated his method to Euler, who recognized its enormous importance. At the same time, Lagrange drew his attention to the principle of least action, enunciated for the first time in a vague way by Maupertuis (1698–1759) in 1744, and, in 1759, he informed Euler that he had almost finished a treatise concerning his method of variation and deduction of mechanics starting with this principle.[16] Euler apparently showed no interest in the treatise (Fig. 4.5).

Before the publication of "Analytical Mechanics", Lagrange published two memoirs in Turin in 1760: "Essai d′une nouvelle method pour determiner les *maxima* et les *minima* des formules integrals indefinies" and "Application de la methode exposée dans le mémoire precedent a la solution de differents problems de Dynamique". Also in 1760, Euler published *Theoria motus corporum solidorum rigidorum,* which was reviewed and extended by his son Johann Albrecht Euler (1734–1800).[17]

[16] Besides d´Alembert's *Treatise of Dynamics*, we already refer to it in (d´Alembert 1921) in Chap. 2. With respect to Euler, in 1744, he wrote *Methodus inveniendi lines curvas maximi minimi proprietate gandentes*, the appendix no. II of which with the title *De moto projectorum in medio non resistente per methodum maximorum ac minimorum determinande*, in which he affirms: *in the trajectories described by bodies under the action of central forces, the integral of the velocity multiplied by curve element, is a maximum or a minimum.* This work is of fundamental importance to variational calculus, which would be developed later by Lagrange using this as one of its starting points.

[17] In 1752, Euler published: *The discovering of a new mechanical principle*, which precisely states Newton's second law in a well known differential form.

In his studies on the rigid body, Euler defined a center of mass or a center of inertia, a consequence of the development of the inertia concept. The definitions of moments of inertia, the decomposition of motion of a free rigid body into the motion of its center of inertia and the rotation around an axis passing through that center also appear. Yet, in the same study, he proposed, for the first time, the classical differential equations governing rigid body motion around a fixed point, with the moment of applied forces to the body, the components of instantaneous rotation and its derivatives together with the moments of inertia around the fixed point also appearing.

These are the main achievements before the formalization performed by Lagrange, which we should mention in order to identify Lagrange's contribution properly.[18]

Lagrange's "Analytical Mechanics" was published in 1788, crowning a series of Lagrange works and other important contributions previously developed by d′Alembert (1717–1783) and Euler in Chap. 1. In relation to the new mechanics, Lagrange stated: *We already have various treatises on mechanics, but the plan of this one is completely new. I propose to reduce the theory of this science, and the art of solving problems related through to formulae, by means of general formulae, and in this way simple development gives all the equations necessary for the solution of each problem. I hope that the method which I have just developed will achieve this objective, not leaving anything wanting.* And he continued: *Figures cannot be found in any part of this book. The methods outlined here do not need constructions, nor geometrical or mechanical reasoning, but rather only algebraic operations, subject to a regular and uniform pace. Those who love analysis will, with pleasure, see mechanics as a new branch, and will be grateful to me for thus having extended its domain.*

4.3 Laplacian Project and Rational Mechanics

Around 1800, Pierre Simon Laplace and Lagrange were the two most relevant figures in French science. If we look at Laplace's work, it is possible to verify that mathematical analysis is applied in two different directions: the celestial mechanics and the theory of probabilities. In 1796, he published *L'Éxposition*

[18] From Lagrange's point of view, the relationship of his investigation to Euler's works can be found in a letter written in 1755 by the young Lagrange, at that time nineteen years old and addressed to Euler, in which he explains that he had derived a general method for treating the problems proposed by Euler in his classic treatise of 1744, that is the *Methodus inveniendi*, which we had referred to in footnote (16). Also important to the further development of Lagrange's work are two memoirs which are known as Turin's memoirs and published in 1760–1761: *Essai d'une nouvelle method pour determiner les maxima et les minima des formules indefinies* and *Application de la method exposée dans le mémoire precedent à la solution de different problems de dynamique.* (See Fraser 1990), p. 198.

du système du monde, a kind of scientific popularization book, without mathematical formalism, where practically all astronomical knowledge of his time is explained. Yet, in this book is found a theory of the appearance and evolution of the solar system, starting from nebulas in rotation. In the domain of celestial mechanics, Laplace explained the fundamentals of his contribution in a monumental book: *Traité de la mécanique céleste*. It first appeared in 1799 (Dalmenico 1992).

With respect the latter above-mentioned book, Laplace studied such important problems as the three body problem, the stability of the solar system, the motion of comets, the moon's theory, the satellites of Jupiter, and others. In addition, he developed specific mathematical tools for application in those studies.

In a comparison between the works of Laplace and Lagrange, one can say that the former was concerned with the application of mathematics mainly for solving problems besides the possibility of finding a general method of analysis as a starting point of each problem. Lagrange, on the other hand, was concerned with the analysis itself from the point of view of the elegance of analytical solutions, as well as in finding a new synthesis of knowledge.

At the turn of the eighteenth century and the beginning of the nineteenth century, as we have seen in the last chapter, a new revolution in physics and in chemical processes occured. Laplace and Lavoisier worked together on heat phenomena and published a research in 1783 called *Memoire sur la chaleur*. At this time, two hypotheses about heat were in dispute: heat as a fluid, the caloric, versus heat as a consequence of motion of particles of matter. In this context, under the strong influence of Newtonian mechanics, the *Laplacian project* then arose.

Thus, in the first decades of the nineteenth century, Laplace and some of his followers began to espouse and attempted to unify the idea that optics refraction, solid cohesion, the effect of capillarity and chemical reactions are the result of attractive forces exerted by particles of matter in a similar way to Newton's universal gravitation. In spite of these propositions not being original, because Clairaut had tried to develop a mathematical theory based on molecular forces in his optics refraction study and in capillarity action between 1730 and 1740, Laplace also studied these two phenomena in his *Traité de mécanique céleste,* published in 1808. He emphasized the universality of this explanatory program and tried to generalize to other physical phenomena, inaugurating the *Laplacian project.*

Laplace rejected the tradition of rational mechanics in favor of a new universal physics based on the hypothesis of molecular motions and forces directly associated with the possibility of applying this new vision to optics and thermal and electrical phenomena.

The *Laplacian project* was presented in a more developed form by Dennis Poisson (1781–1840) as a theory of *physical mechanics* with the objective of replacing Lagrange's *analytical mechanics.*

Poisson defended this new mechanics, emphasizing that molecular forces could be applied to similar problems solved by Lagrangian mechanics, such as vibrating strings, elastic surfaces and fluid mechanics.[19]

In spite of the failure of the *Laplacian project* in trying to unify a series of physical phenomena under the aegis of mechanics, it had significant importance to the development of physics throughout the nineteenth century. It had encouraged experimental works, to privilege the measurements and experimental methods as a paradigm of the sciences of the physical world.

And yet, under the influence of Laplace's ideas, his follower and associate, the chemist Louis Berthollet (1748–1822), postulated that chemical affinity was also the result of attractive forces among particles of matter and stated that this phenomenon and gravitational attraction were quite similar. He emphasized the importance of mass to chemical affinities. Despite the difficulties and even the failure to establish an explanatory framework for chemical reactions based on experimental results guided by these ideas, the importance of this vision to further developments of this science was relevant.

Obviously, the *Laplacian project* failed to provide a theory in order to unify the particular sciences of the physical world during the first decades of the nineteenth century, although its collapse was accelerated by political questions. During the Napoleonic period in France, it was supported by the government that disappeared with Napoleon's fall in 1815. Thus, the theory of affinities was replaced by the atomistic theory of Dalton, the theory of caloric of Fourier's studies, and later by thermodynamics, the cornerstone of Laplacian theory, and the corpuscular optics of the wave theory of light. At last, the mathematization of the physical world, the experimental methods and quantification, alongside the representation of the physical world by models, were a rich heritage incorporated into physics in further developments.

With the decline of the *Laplacian project*, mechanics were now hegemonically represented by Lagrangian mechanics, which became a paradigm of physical world sciences spreading their influence to social sciences and the economy. Along the way, the ideas and the Cartesian project of mechanics, a well-formalized theory without experimentation that was derived purely from reason, became dominant. It was satisfactory to them but, on the other hand, created a serious problem by revealing the incapacity of rational mechanics to formulate a theory to explain machines within the framework of rational mechanics. In addition, the context of an explosive capitalist expansion, mainly in England where a machinist

[19] In fact, the well-known Laplacian project was sponsored by a significant number of young researchers. They founded a kind of center of investigation known as *Societé d'Arcueil.* Briefly, the history of this group began with the return of Claude-Louis Bertholet from a campaign in Egypt in 1799. He bought a property in Arcueil, a village located five kilometers to the south of Paris. Laplace and Bertholet had already been friends since the previous election to the Sciences Academy in 1780. Both were close to Lavoisier and active participants in French science. Other young researchers, such as Jean-Baptiste Biot, Louis-Jacques Thénard, Joseph Gay-Lussac, and others, joined them. After 1807, Etienne Louis Malus, Dominique François Jean Arago and Simeon Denis Poisson also joined the group. See (Fraser 1990).

expansion had made astonishing progress, aggravated the dilemma. The solution of this dilemma began to be solved by Lazare Carnot and his followers, who were the polytechnician engineers of the three first decades of the nineteenth century. This history will be studied in Chaps. 5 and 6. Before this subject, we will look briefly at the the development of mechanics as used by craftsmen and machine and mechanism constructors, as well as the diffusion of empirical mechanical knowledge.

4.4 Empirical Knowledge of Machine Manufacturers

In the preface of *Principia*, Newton wrote: *The ancients considered mechanics in a twofold respect; as rational, which proceeds accurately by demonstration, and practical. To practical mechanics all the manual arts belong, from which mechanics took its name. But as artificers do not work with perfect accuracy, it comes to pass that mechanics is so distinguished from geometry that what is perfectly accurate is called geometrical; what is less so, is called mechanical* (Newton 2002) in Chap. 2. As Newton remarked, there are two classes of mechanics. Throughout their history, both followed parallel paths but sometimes these paths would cross each other with resulting interchange. Some important figures of theoretical mechanics dedicated portions of their work to the construction of a varied sort of machine and mechanism.[20]

We begin our history with the contribution of the School of Alexandria. This city was founded by Alexander the Great, in November of 332 BC, when he arrived in Egypt. It became the second most important city in Egypt and also an important sea port. During this period of Greek history, Alexandria became a center of intense commerce between Europe and the Orient and, in a period of less than one century of existence, it overcame Cartago in size. The relevance of the city as a Hellenistic center began to appear as it became a privileged place for the development of Greek science. Alexandria's library accumulated 500 thousand volumes and other institutions were created to be museum and an astronomical observatory. Thus, Alexandria became a complex center for scientific investigation.[21]

[20] Some cases are well known, such as Galileo, who was an expert device constructor, the case of telescope perhaps being more well-known. The second example is Huygens, who, as we have seen, had an essential contribution to the improvement of the precision watch. The third case is Leibniz, who built a machine to calculate undertaking the four operations. However, he was a mining engineer in Harz, a place far from Hanover by about one hundred kilometers. He designed windmills rotating on a horizontal plane instead of a vertical plane as was usual at that time.

[21] The School of Alexandria's contribution to science is extremely important. In mathematics, we have Euclid, who directed Alexandria's museum between 330 and 221 BC. Apollonius of Perga, a city located in the south of what is now Turkey, worked in Alexandria between 246 and 221 BC. He was author of the book: *On the conics*. In mathematics and mechanics, we have Archimedes (287–212 BC), Hero and Pappus. In astronomy, another scientific branch which occupies a remarkable position in that school, we have Aristarco, who was born in Samos and lived between 310 and 230 BC. Also, Hipparchus from Nicaea and Ptolemy, who was born in Egypt around the year 100 of our era. He spent his lifetime in Alexandria where he died around the year 170.

With respect to studies of mechanics applied to machines, Alexandria achieved great progress. The primary findings and developments are associated with Archytas (428–347 BC), Ctesibius (285–222 BC), Philon of Byzantium (20 BC–50 AD) and Hero of Alexandria (10–70 AD), with the latter being the great constructor of automata (Koetsier and Blauwendraat 2004). In 1693, four of Hero's works were published: *Treatise on Machines of War*, *Pneumatics*, *Automata* and *Chirobalistics*.

The contribution of the Alexandria School to applied mechanics is remarkable. Besides the scientific advances, various types of equipment and devices were constructed: military equipment, measurement instruments, toys and a lot of domestic devices. Since the end of the fourth century BC, the simplest machines were the pulley, the lever, the screw and the wedge, with general utilization. The screw was then improved to be used as a water pump. This type of device is known as Archimedes' screw .[22]

A series of inventions are associated with Ctesibius. It is only known that he worked in Alexandria around 270 BC and that the devices constructed by him come from his investigations into compressed air based in pneumatic principles. Two of these inventions were a compressed air pump with valves and a hydraulic pump. He also designed military equipment moved by springs and compressed air. Other famous creations were clepsydras operated by water flow with levers and automatic pieces such as bells, automats and birds. He is known as inventor of the precursor of the cuckoo clock.

The contribution of Hero as a machine constructor is the best known. The publication of *Automata* caused a great deal of interest within the technical community. He is universally known as the inventor of a steam machine and a gear train. This "machine" is, in fact, a toy that is demonstrated on the first page of thermodynamics textbooks. However, it seems that Hero has been popularized by his constructions of automata. He demonstrated a great ability in these constructions by applying modern principles to make the mechanisms move. If we analyze principles of design, it is easy to realize two basic ideas behind them: a program and a regulation. A kind of program appears in the design of chariots because the inversion of the motion from one sense to another clearly follows a program. The construction of the mechanisms for transmission of motion with pulleys, strings and gear trains is remarkable.

The difference between Hero and Ctsibius does not appear in their individual skills but in the physical phenomena used. Hero systematically uses air compressibility and water incompressibility. This is done without a theoretical formulation about the concepts. He applied these principles in his two most popular automata: the mechanism for opening the doors of a temple and the thermal turbine. Ultimately, Hero constructed many other machines and devices with a military purpose. This kind of motivation was a characteristic of this period, creating a favorable environment for the innovative process. This situation would only be repeated in the Renaissance.

---

[22] Historians and experts believe that is fair to assume that Archimedes invented both the helix and the screw-pump. Archimedes, with these inventions, was interested in the problem of squaring the circle. A great number of ancient texts refer to Archimedes' pump, attributing its invention to him. See (Koetsier and Blauwendraat 2004).

In order to finalize this summary of the Alexandria School period, we should mention Archimedes' contribution to the design and construction of machines. Archimedes probably spent a period of his youth in Alexandria for study. Thus, he had the opportunity to know mathematicians such as Conon of Samos and Eratosthenes of Cyrene. In general, the literature of the history of sciences only mentions his contributions to mechanics, such as the principle of the lever and hydrostatics. However, Archimedes designed many machines by assembling mechanisms with pulleys in cranes of different sizes. Some of them were used in the defense of Siracusa, many others for moving ships or even destroying them. With respect to military machines, Archimedes designed and constructed catapults for launching arrows over long distances, a capstan to develop tension in an arch in order to accumulate energy for launching, platforms to feed arrows to the war machines, and a kind of a riffle.

During the Arabian period, which followed immediately after the Alexandrian School, other interesting advances worth highlighting took place. Many of these were the work of Al-Jazari (1136–1206), an engineer and constructor of machines with much in common with Hero. Al-Jazari left a work published in 1206 entitled *Treatise on the Theory and Practice of the Mechanical Arts*. It is a type of theoretical manual describing machines which can actually be constructed and gives, in addition to an explicative text, drawings and plans to guide in their construction. In spite of difficulties in evaluating the influence of this work in the further development of machines, it is possible to identify certain similarities in more recent years with the machines described by Al-Jazari. He followed the tradition which came from Alexandria concerning fountains, clepsydras and systems of power transmission. In addition, he worked on machines for pumping and storing water.

A new mechanism appeared in this period: the rod-crank which would be introduced into Europe only three centuries later.[23] Obviously it was not a modern system, but it was a very innovative one. It was the first attempt to transform a continuous circular motion into an alternative motion through a sliding rod.

Al-Jazari also contributed to the construction and maintenance of clepsydras, a real concern of Arab engineers. He made some improvements in clepsydras, such as calibration of orifices, utilization of paper models in understanding complex drawings, construction of wood gages, and a study of the static balance of disks. He made innovations in system regulation and feedback with the clear influence of Ctesibius.

With the fall of Bagdad in 1,258, this technical culture was forgotten for a long time.[24]

---

[23] To obtain an explanation about the rise of the crank-shaft system, see Bertrand Gille's paper in (Gama 1985), p. 173. According to Gille, the first image of a crank-shaft system appeared in a German manuscript from Munich in the period 1421 to 1434. It is a windmill moved by an arm. Leonardo da Vinci reinvented the mechanism, but its practical construction had problems to overcome.

[24] In this period, Bagdad was a great trade center and an important cultural city. Bagdad's fall in 1258 was due to the Mongol invasion of Mesopotamia, commanded by Hulagu, who pillaged the city, killed the caliph Al- Mustasim and, according to some sources, killed eight thousand other people as well. Bagdad never resettled completely. Only with the emergence of Iraq as an independent state after the Second World War did Bagdad receive a new impulse for development.

Fig. 4.6 Water wheel (Ramelli's book)

Due to the repercussions they would have on industrial processes, the most important machines from the Arabian period to the Middle Ages were water wheels and watches. In Chap. 2, we briefly described the history of the watch. Now, after the history of the water wheel, we focus our attention on steam engines, which would become one of the characteristics of the first phase of the Industrial Revolution in England (Fig. 4.6).

The water wheel, ancestor of the hydraulic turbine, made rapid progress from the eleventh century onwards due to its importance for the economic system of Western Europe. From the technical point of view, these water wheels did not differ much from those used by the Greeks and Romans several centuries before. Their importance was due more to the fact that they functioned as an essential part of a productive unit, located in the field or in the city.[25]

The classic work on the study of the origins of the water wheel is Vitruvius' treatise called *On Architecture*, published in the first century BC. This machine is contemporary to ancient Greek and Roman civilizations. Its rise in the Mediterranean region is practically simultaneous to its appearance in China. There are documented sources of these machines in Italy, Portugal, and the Scandinavian

[25] Lynn White Jr., in his paper *Technology and innovation in middle ages,* p. 88 of (Gama 1985), affirms: *Since the eleventh and twelfth centuries a great substitution of human to non-human energy occurs in any place where big amount of energy and force is required or where the required motion can be replaced by a mechanism. The late Middle Age glory was not the construction of cathedrals, his epics or his scholastics; it was for the first time in history the construction of a complex civilization supported not by the slaves, but by non-human energy.*

countries, as well as France and even a small number of unities before the ninth century. One century later, a great expansion of these machines came about due to economic changes such as agricultural transformations, exploration of forests and progress in metallurgy, creating new demands for iron products.

In spite of its ancient origin and large application, the water wheel would only be scientifically and systematically studied at the beginning of the nineteenth century. However, in the eighteenth century, Bélidor conducted a pioneering study about this system. In the previous centuries, innovations and improvements were made by machine constructors and craftsmen replacing damaged pieces.

In the transition from the Middle Ages to the Renaissance, two figures are of great importance to the development of hydraulic wheels: Francesco di Giorgio Martini (1439–1502) and Leonardo da Vinci (1452–1519). The first left a great number of hydraulic wheel drawings which are a clear anticipation of a hydraulic turbine. Leonardo was concerned with these machines and studied water motion to observe the formation of vortices. During the years from around 1508–1510, he designs various hydraulic machines, obtaining interesting results. These studies are contained in the Codex Leicester and in Manuscript F.

As we mentioned at the beginning of this section, in some cases the theoretical production and movement of practical men have met. They also influenced each other. During the fifteenth and sixteenth centuries, a kind of movement of practical men and machine constructors arose fighting for social recognition. Many manifestations of this socio-cultural movement appear as literary works of a particular form. This is, for instance, the case of Agricola (1494–1555). His masterpiece *De re metalica*, had the purpose of gaining recognition for the less-considered professions in society. This book written in Latin, the language of literates and erudites, is a strong sign of this claim. Other important figures did the same, such as Alberti, Piero della Francesca, Martini, Biringucio and Tartaglia.

At the end of the sixteenth century, practical men and machine constructors had criticized the official theories and their adherence by professionals. This appears in the work of Vesalio in his book *De humani corporis*, published in 1543. It is a critique of doctors with respect to their disdain for manual work.

This brief parenthesis is to emphasize the importance of recognition of practical knowledge for a genuine Scientific Revolution. The introduction of the experimental method was significantly influenced by this movement and represented a new form for ordination of the existing knowledge. The experimentation also meant generalization and abstraction. Consequently the progress of practical knowledge created favorable conditions for the rise of figures of Galileo's magnitude.

In relation to the steam engine, it emerged from the experiments of Otto von Guerick (1602–1686) carried out in 1654.[26] Nonetheless, its origins are related to

---

[26] In (Lilley 1965), p. 99, one reads: *In 1650, Otto von Guericke, from Magdeburg, invented the pneumatic pump, which creates the vacuum and demonstrates with the help of huge forces what atmospheric pressure could provide to man; in one of these experiences, the created vacuum in a small reservoir originated forces such that eight horses cannot separated.*

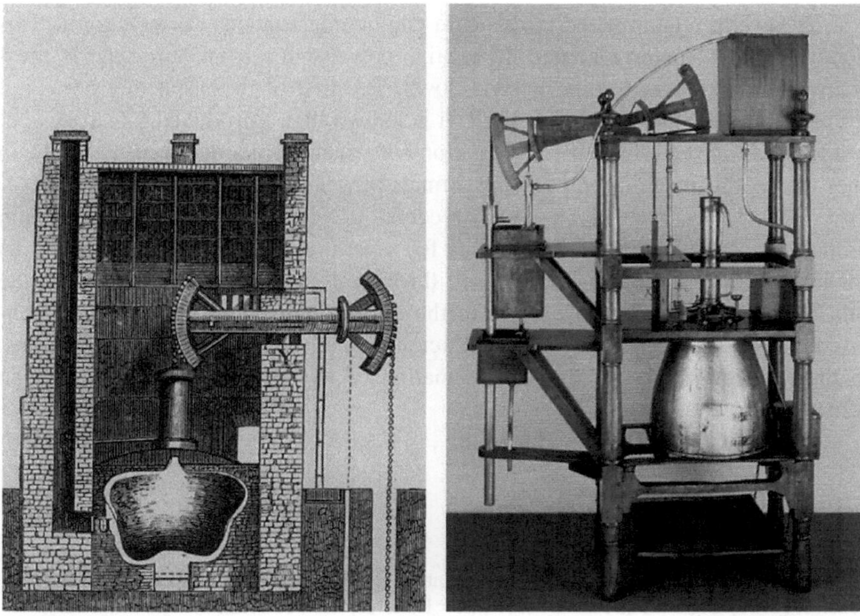

Fig. 4.7 Newcomen's machine and a model

Fig. 4.8 Watt machine

the works of scientists and machine operators in the seventeenth century, such as Huygens (1629–1695), Denis Papin (1647–1712), Leibniz (1646–1716) (Lilley 1965), and others who systematically worked with the problem of atmospheric pressure and the use of heat as a form of energy. Papin tried to construct an atmospheric machine in 1707, with the aim of pumping water into a reservoir. Thomas Savery (1650–1715) built a machine with the same purpose in 1698.

After several attempts, the third steam engine constructed was successful. It is known as Newcomen's machine. Its motive power came from atmospheric pressure while a piston was being pushed down. The first of these machines was put into operation in Staffordshire in 1712. It is known that it produced 12 strokes per minute, each one of which lifted 10 gallons of water from a depth of 51 yards, in other words, this machine had an approximate power of 5.5 hp (Fig. 4.7).

In 1763, James Watt (1736–1819) received a Newcomen (1664–1729) steam engine to repair. He worked on the idea for several months and then modified its structure so that the steam used passed through a condenser separated from the main cylinder. Watt's improvements to the steam engine in 1769 and 1784 converted a device of limited use into an efficient machine with multiple applications. He thereby contributed to the transformation of an agricultural society into an industrial one (Fig. 4.8).[27]

References

Blay M (1992) La Naissance de la Mécanique Analytique, PressesUniversitaires de France, Paris
Brenan R (1998) Gigantes da Física. Jorge Zahar Editores, Rio de Janeiro
Corbisier, R (1988) Introdução á Filosofia, Tomo II, Parte Segunda. Editora Civilização Brasileira, Rio de Janeiro
D´Alembert JL (1921) Traité de Dynamique. 2 Vol, Gauthier-Villars et Cie. Éditions, Paris
Dalmenico AD (1992) Mathematisations: Augustin-Louis Cauchy et lÉcole Française, Éditions du Choix. Librairie Scientifique Albert Blanchard, Paris
Dalmenico AD, Pfeiffer J (1986) Une Histoire des Mathematiques. Éditions du Seuil, Paris
Dugas R (1988) A history of mechanics. Dover Publications, Inc., New York
Einstein A (1999) A teoria da relatividade especial e geral. Contraponto, Rio de Janeiro
Eves H (1997) Introdução à História da Matemática, 2ª. Edição, Editora da Unicamp, Campinas, S. Paulo
Fraser CG (1990) Lagrange's Analytical Mathematics, its Cartesian Origins and Reception in Comte's Positive Philosophy. Stud Hist Phil Sci 21(2):243–256
Gama R (1985) História da Técnica e da Tecnologia. Editora da Universidade de S. Paulo, S. Paulo
Geymonat L (1997) Galileu Galilei. Editora Nova Fronteira, Rio de Janeiro
Koetsier T, Blauwendraat H (2004) The archimedean screw-pump: a note on its invention and development of the theory. In: HMM2004 symposium, Cassino, Italy

[27] In fact, the question of the social impact caused by the steam machine and the subsequent machinist expansion is a little bit more complex. Michel Vadé, in a recent biography of Marx (Vadée 1992), discusses this question as follows: *The development of modern machinery seems to be provoked and determined by steam machine. Several kinds of machines only appear after its appearing as a direct or indirect consequence. Is indeed from the steam machine that depends the significant growing and the possibilities of the production of the technical industrial system? Marx postulated a contrary point of view facing this current idea. His thesis is a kind of paradox, according to which the Industrial Revolution was not caused by the sudden quantitative growth in energetic power developed by steam machine and its continuous improvements, but by the qualitative revolution within the machinist itself, deeper and more radical then the invention of the steam machine.*

Kuhn TS (1957) A Revolução Copernicana, Edições 70. Ltda, Lisboa

La Mason SF (1986) Ciencia Del Siglo XX. Aliança Editorial Madrid, Espanha

Leibniz GG (1983) Oeuvre concernant le calcul infinitesimal. Trans. latin to french by Jean Peyroux, Librairie Albert Blanchard, Paris

Lilley S (1965) Hombres, Maquinas y Historia. Editorial Ciencia Nueva S. L., Madrid

Newton I (1994) La Méthode des Fluxions et des Suites Infinies, translated to French by M. Buffon. Scientific Library and Technics Albert Blanchard, Paris

Newton I (2002) Principia: mathematical principle of natural philosophy. Book I, Edusp, University of S. Paulo Editors, S. Paulo

Poincaré H (1970) La Valeur de la science. Flammarion, Paris

Ronan CAA (1987) Ciência do Século XX, IV Volume de História Ilustrada da Ciência. Universidade de Cambridge, Cambridge

Russel B (1967) História da filosofia ocidental, vol 3. Companhia Editora Nacional, S. Paulo

Truesdell C (1983) Essays in the History of Mechanics, Springer-Verlag Editors, Berlin

Vadée M (1992) Marx Penseur Du Possible. Meridean Klinchsieck, Paris

Part II
The Instrumental Genesis

Chapter 5
Lazare Carnot's General Theory of Machines

If the principle of inertia force, of composed motion, and of equilibrium, are essentially different one of the other, as we cannot prohibit to happen; and if on the other hand these three principles are sufficient to mechanics, one can reduces this science to the least number of principle possible, and assume that it is established on these three principles all bodies' law of motion for any circumstances, as I have accomplished in this work.

(Jean Le Rond d'Alembert—Preliminary Discourse to *Dynamics Treatise*, G. Villars et Cia., Editors, Paris, 1921.)

5.1 Lazare Carnot: Scientific and Political Career

Lazare-Nicolas-Marguerite Carnot was born on May 13th, 1753 into a bourgeois family from Bourgogne in France, which occupied a remarkable position in local society. Claude Carnot, his father, was a lawyer and public notetaker in Nolay, a small city close to the Côte-d'Or. The place where Lazare Carnot was born has not changed significantly over time and still belongs to Carnot's family to this day (Fig. 5.1).

Carnot's father took charge of the education of his three older children and they, in turn, helped the younger brothers as tutors or preceptors. Among the six sons, two demonstrated an early ability for mathematics and technical questions. These characteristics meant that Lazare, the second son, and the younger Claude-Marie (1755–1836), soon found themselves on the path to careers as military engineers.

During the 18th century, in order to assure a son's education it was customary, particularly amongst the wealthier families, to ask for the help of a priest or a preceptor. There were only two options for such an education: classic or humanistic studies or military schools. At the Universities, one could engage in courses in the law, medicine and theology.

A. R. E. Oliveira, *A History of the Work Concept*, History of Mechanism and Machine Science 24, DOI: 10.1007/978-94-007-7705-7_5,

Fig. 5.1 Lazare Carnot (1753–1823)

Carnot completed his humanistic course at the Autumn School, which was taken over by oratories after the Jesuits had left the school. Ten years later, Napoleon Bonaparte studied there. Lazare Carnot finished these studies at the age of 16.

With the idea saving his father's money, Carnot decided to study alone before submitting to the examinations at the Mézières School of Engineering. However, his attempt to be admitted to that School at the end of 1769 proved unsuccessful, so his father went with him to Paris. In addition, Carnot obtained the patronage of the Duke of Aumont and thus he was admitted to one of Paris's most famous institutions run under the direction of Louis-Siméon de Longpré and located in the historic Le Marais district. Longpré had long supported himself by giving instruction in mathematics, and soon after became an author. Being well versed in mathematics, he conceived the plan of erecting a school for the preparation of young men destined for the army.

Near the beginning of his time there, Carnot had the opportunity to meet Charles Bossut (1730–1814) and d'Alembert, two figures of great influence on Carnot's approach mechanical problems. Bossut, who directed Mézières's examinations, published, in 1763, the first edition of a very popular mechanics handbook, entitled *Elementary Mechanical Treatise and Applied Dynamics mainly to Machines Motions.* He also published other algebraic, geometrical and arithmetic works.[1] Alongside his primary interest in mathematics, he was particularly concerned with

[1] It is also very popular his *Essais sur l'histoire génerale des mathematiques*, in two volumes.

mathematical education. He wrote *A General History of Mathematics from the Earliest Times to the Middle of the eighteenth Century*, which was published in 1802.

At that time d′Alembert was an old man and maintained a strong friendship with Longpré. The great mathematician frequently met with the students around him to suggest mathematical problems. Later on, when Carnot was also an old man, he enjoyed telling how d'Alembert had influenced his method for solving difficult questions and had predicted a successful career for him.

After securing approval through his examinations, Carnot entered the Mézières's School of Engineering, in January of 1771, and began the 2 years of basic formation which were fundamental to a future career in engineering. At that time, the Engineering School required proof from candidates that they belonged to the noble class or at least to a family living in a noble manner. In other words, they had to show that they came from a life free of degradation. It is important to note that the majority of the students were of bourgeois origin and Carnot had no difficulty obtaining the certificate.

Unlike Carnot, Gaspar Monge (1746–1818), one of the greatest talents among that group of students, suffered cumbersome obstacles because of these requirements. Monge's family had modest resources and so he had to rely solely on his intellectual qualities to gain admittance. Fortunately, Monge was able to overcome all these difficulties and became a lecturer in mathematics and physics.

Concerning Monge's influences on Carnot, there are divergences among biographers. In fact, none of the biographers refer to each other's work and there is no clear indication that Monge was interested in Carnot's work. There were also profound differences of personality between the two men. Monge was a mathematician and a pedagogue, without any tendency towards politics. When he did become involved in political issues, he demonstrated a theoretical spirit, being very emotional and incapable of making decisions, besides displaying no sureness for judgment and no attention to details. Carnot, on the other hand, behaved inversely, being an engineer with clear and objective ideas, the ability to conceive of and execute his tasks with uniformity of character, and preferring to judge with a practical spirit and adaptation to circumstances.

Carnot left Mézières on 1 January 1773, beginning his military career by serving at several military posts, initially in Calais, and then in Havre, Béthume, Arras and Aire. In Calais he worked on fortification of the port. Later on, he moved to Havre and worked there for 3 years in order to participate in the building of the port of Cherbourg, an advanced technical project considered one of the most modern in French military engineering. With this and participation in other projects, Carnot became considered among his peers a well-known and reputable engineer.

In order to understand Carnot's scientific work, perhaps we should remember Coulomb's quotation from 1776, concerning the situation of all graduates of engineering school. He states: *a young men who studied and after left the School had no other thing to do, to face tedious and monotony into the corporative situation that is the dedication to any science branch or to literature absolutely strange to his study.*

Obviously, this is what happened with Carnot after he finished the course at the Mézières′s School. He spent part of his time in mechanical and mathematical studies, achieving a higher level of knowledge when compared with other students from Mézières. His courses had provided the study of four books of Charles-Etienne-Louis Camus[2] on arithmetic, geometry and statics, as well as Bossut's treatises on dynamics and hydrodynamics; as a compliment, there were annual courses on industrial design, perspective or descriptive geometry, as well as an experimental physics course. Camus is an important French mathematician and mechanician, who also studied civil and military architecture and astronomy. He became secretary of the Academy of Architecture and a fellow of the Royal Society of London.

When Carnot left Mézières's School, facing the solitude of life at military posts, he likely began to re-read d'Alembert, Bossut, Bélidor, Euler's mechanics and Daniel Bernoulli's hydrodynamics. This is an easy conclusion when we read his early works, in which he demonstrates a good knowledge of the above subjects. Another argument to corroborate this supposition is his distance from the scientific circles in a different way, followed by Coulomb and Meusnier. Sometimes Carnot spent his vacations in Nolay or Dijon. The documents from the Paris Sciences Academy and the German Sciences Academy also confirm this supposition. Other manuscripts belonging to Carnot's family reproduce in some way the circumstances and the context under which he wrote the early works on mechanics and mathematics.

The first of Carnot's studies were addressed to his participation in concourses sponsored by scientific societies at that time. In 1777, the Paris Academy promoted a concourse for the year of 1779, but it was only accomplished in 1791, because the referees were not satisfied with the works submitted the first time around. Consequently, Carnot wrote two of his first studies on machine theory. Both entitled *Essay on machines*, he concluded the first in Cherbourg in March 1778 and the second in July 1780 in Béthume. Coulomb's memoir prevailed as the winner while Carnot received an honorable mention.[3]

With respect to his interest in mathematical foundations, the motivation is quite similar. It is well known from his text *Considerations about the metaphysics of infinitesimal calculus*,[4] a memoir submitted to the Prussian Royal Academy in 1785. Then, following the concourses' motivations, the Paris Academy asked for anyone interested to write about the first human flight. Carnot met this need on January 17th, 1784 with *Letter on aerostats*. We should remember that on June 5th, 1783, the Mongolfier brothers launched a hot-air balloon which achieved a height of 1800 m. This event encouraged the solution of several technical problems concerning the motion and stability of flight.

[2] The most known Camus's works are: *Élements de géometrie théorique et pratique* (1750), *Élements de mécanique statique* (1751) and *Élements d'arithmetique* (1753).

[3] Coulomb's memoir entitled *Théorie de machines simples* will be analyzed in depth in the next chapter.

[4] This work adopts the Leibnizian approach to investigating differential calculus and tries through the theory of limits to eliminate some conceptual inconsistencies and improve its algorithmical aspects, which was a Leibnizian characteristic.

Carnot studied the first the air-ship problem and proposed a very unusual, fantastic solution in the form of a propeller which displaced itself like a medusa working according a systole and diastole motion created inside the balloon through heat dissipation.

These are some of the most remarkable facts related with the beginning of Carnot′s scientific career. In the following pages, we will see his participation in politics primarily at the beginning of French Revolution.

If we need to characterize Carnot and form a true picture of his character, we must begin by calling him an authentic republican. He was involved with important circumstances of the French revolutionary period, for instance, voting for the death of Louis XVI, confirming his republican convictions. On the other hand, his military position contributed to his belief that the leftist movements were the main menace to the recently created Republic. Hence, he fought against the revolutionary movement known as the *Conspiracy of Equals,* commanded by Gracchus Babeuf (1760–1797).

Babeuf‘s name is a tribute to the Gracchi brothers, Roman tribunes of the people and reformers. He was a French political revolutionary and a journalist of the Revolutionary period. In spite of the efforts of his Jacobin friends to save his life, Babeuf was arrested, tried and convicted for his participation in the *Conspiracy of Equals.* At that time, the words anarchist and communist did not exist. *Conspiracy of Equals* was an armed rising fixed for Floréal 22, year IV (11 May 1796). One day before Babeuf was arrested, many of his associates were gathered by the police by order of Lazare Carnot. Babeuf was guillotined at Vêndome Prairal 8 (27 May 1797) without appeal. This episode is important from the viewpoint of revolutionary movements since the French Revolution because some historians see this particular rebellion as the first proletarian movement within bourgeois revolutions.[5]

Carnot also participated in regrouping the French army after the disorganization caused by the Thermidorian movement by maintaining control over the military administration. However, the French Revolution had plenty of surprises, with victories and defeats of the republican forces and movements for reestablishing the monarchy.

In spring of 1797, the legislative elections showed the possibility of a resurrection of the monarchy. The left could not accept the defeat of the Revolution and thus tried to reject the results of the elections, as well as attempting to eliminate the reactionary members from the assemblies. In the *coup d'état* that occurred on

[5] The Brazilian historian Carlos Guilherme Mota analyzes this movement and affirms: *It is the radical proposal to a new society which births from the interior of Revolution.* He continues: *With the takeover of power, it doesn't make sense to propose a Democratic and Liberal Assembly, but to organize the revolutionary dictatorship for the necessary time to implement the new society.* As we know, the movement was defeated after being denounced to Carnot by a militant whose name was Grisel. Carnot, in the position of a strong man of the regime, appointed Cochon de Lapparent on April 3rd of 1796 as the minister of police, a famous enemy of the Jacobins, which commenced the struggle against the "communists". See (Mota 1989), p. 183.

18 Frutidor (September, 4th), Carnot was convinced that was not possible to maintain the Constitution, and did not participate in this movement. Receiving communication concerning the directory's intention, he left Luxembourg and took refuge in Paris for some weeks before moving on to Switzerland.

In the midst of this political situation, Carnot frequently spent his time in researching mechanics and mathematics, according to his son Hippolyte. In a similar situation in 1797, during his participation as a member of the directory, his book *Considerations about the metaphysics of infinitesimal calculus* was published.

With the taking of power by Napoleon Bonaparte in 1799 (18th brumaire), Carnot came back to France by benefit of a general amnesty in favor of Termidor's victims. He soon relinquished his position after realizing that this new regime was creating difficulties in his own affairs, aside from their hostility to the Republic. At that time, certain works were published. In 1800, at the end of the year, he published his *Letter to Bossut citizen containing new visions about trigonometry*; in 1801, *From the correlation of the geometry's figures;* in 1803, *Essay on machines* with the new title of *Fundamental principles of equilibrium and motion*, which occupies a central position for the purpose of our book. That same year, he also published *Geometry of position*, which he personally considered his masterpiece. And in 1806, appeared his *Memoir about the relation that exists among the respective distances of any five points taken in space*, followed by an essay about the theory of transversals.

The two sons of Lazare Carnot, Sadi and Hipolyte, were born in 1796 and 1801, respectively. Sadi was the founder of thermodynamics, and we will come back to him in order to analyze the influences that his father had on his work.[6]

Another scientific activity developed by Carnot must be emphasized but it is necessary to examine it through the lens of the French Revolutionary period in which it occured. At the beginning of the terror, August 8th, 1793, the Sciences Academy, as well as other French institutions representing the *ancient régime*, had been abolished or closed, because they were considered to be founded on privileges and not adequate to the spirit of the Republic. Carnot had only become a member of the Committee of Public Safety on August 7th, so we cannot attribute the decision to close the Sciences Academy to him. As we know, with the end of the Academy, the French Institute was created, with Carnot appointed as a member of this institution, the purpose of which was to give continuity to the old Academy's works. The new institution was intended to unify the arts, letters and science, as branches belonging to the same culture, being considered the ultimate symbol of the Republic.

[6] Lazare Carnot and Sophie Dupont had three sons: their first, also named Sadi, was born in 1794 but soon died in 1796. The second Sadi Carnot, born on July 1st, 1796, was a physician and founder of thermodynamics, while his brother Hippolyte was born on April 6th, 1801 and died on March 13th, 1888. He was a deputy, senator and minister of public instruction, as well as a member of the Institute of France.

In 1805, Carnot became president of the French Institute. His primary activity was working for the progress of science, as well as publishing memoirs and recommending the addition of other studies submitted to him and considered of importance for dissemination, although he sometimes worked as a public consultant analyzing and judging technological problems, studying the conception and design of new machines and the viability of industrial processes. At that time, it was common for inventors and machine constructors to submit their designs to the government in order to obtain financial resources.

Carnot frequently participated in technical commissions within the attributions of the Institute in order to evaluate projects for their practical characteristics. One example of this was the case of the Niepce motor. In an enthusiastic report, he emphasized that this machine was the first to obtain its primary energy in a way never before imagined, by means of heated air expansion, or in other words, the air combined with the caloric. In contrast with the steam engine, which consumes great amount of caloric to heat and vaporizes the water before using steam expansibility, an air motor presented a terrific advantage in that it does not use combustibility unless producing the expansion where energy arises.

In 1809, Carnot addressed the Institute with a new report he had written about a new model of thermal motor conceived by an inventor named Cagniard de Latour. This motor had attracted the attention of Sadi Carnot, according to Kuhn (1996) in Chap. 1. Like the Niepce machine, the fluid used was the air, but its utilization was more subtle. Some characteristics of this motor are described by Sadi Carnot in his *Considerations*. In addition, Kuhn refers to another important aspect of Cagniard′s motor: its reversibility. This concept appears in Sadi Carnot′s works and we will come back to it later on.

Carnot's final years can be briefly outlined. When Napoleon came back from Elba Island, Carnot allied with him, having been his interior minister in the well-known government of 100 days. However, because Carnot had voted in favor of the death of Louis XVI, further confirming his republican convictions, he could never be forgiven by the restored monarchy. Subsequently, Carnot once again went into exile, with his son Hipolyte. After passing through Brussels, Munich, Wien and Cracow, they arrived in Warsaw where he hoped to stay and live. However, it was not possible for him to stay there, and because of the prohibition stating that he could not live anywhere within the portion of the Rhineland close to France, they ultimately went to Magdeburg, where Lazare Carnot died on August 2nd, 1823.[7]

[7] Andrè Fridberg describes Carnot's death: *Carnot's death seems has been relatively brief. He suffered of digestive perturbations, and did not take care according with his son, besides the advices received of his family. When he agreed to receive a doctor he spent the time for the interview to discuss scientific and political questions without speaking about the disease. Some days later, Lazare woke up early morning, as usually done in other days, made the toilet and shave himself. Getting fatigued, goes to bed, and during the day expired without apparent suffering.* See (Fridberg 1978), p. 221.

Fig. 5.2 **a** Delacroix painting (1830) **b** Declaration of the human rights

5.2 Science and the French Revolution

The great historian of the Russian Revolution, Isaac Deutscher (1907–1967), discussing the meaning of the bourgeois Revolution, said: *The more substantial and permanent result was the elimination of social and political institutions which have increasing the difficulties to bourgeois property to growth besides their social relations. When the puritans denied to the Crown the power to promote an arbitrary taxation, when Cromwell has guarantee to Britain ship-owners a monopolistic position within sea commerce in England in the context of foreign countries and when the Jacobins have abolished the rights and feudal privileges, they created frequently the conditions to manufacturers, to traders and to bankers to achieve economical hegemony as well in the future social and political supremacy. The bourgeois Revolution creates the conditions in which bourgeois property could develop. It is in these aspects and not in particular alignments which appear during the fight that consists its specific difference* (Fig. 5.2).

It is not our intention to analyze the French Revolution, even superficially, but the above citation appears in order to underline its importance as a fundamental transformation of European society from the social, political and scientific viewpoints. Besides the intellectual and cultural changes before the takeover of power by the bourgeoisie in France with direct consequences to the scientific production over a long period of his history, we should also mention other factors and aspects of this context. The first important thing to note is the need of the new regime for new institutions in order to criticize and fight against the ideas of the *ancient régime*. As Deutscher's comment emphasizes, it was necessary within the framework of the new situation to eliminate the ties and the previous difficulties imposed by feudal property in general. These new institutions also appeared within the educational system as the best way to change mentalities and to prepare a new technical and political elite to give continuity to the project of a new society announced by the Revolution.

The transformations in the educational system of France implied significant modifications in technical and professional education, because new creeds, new knowledge and new technologies had progressed, making it necessary to teach them. The development of engineering and its teaching is important in this context. The early origins of engineering in France are found in military activities, mainly in artillery and fortification. In the first half of the sixteenth century, the Italians invented bastion fortifications, which rapid spread rapidly throughout Europe. This important improvement provided an increase in importance of artillery and fortification specialists in military activities. As a result, a *corps* of officers was gradually organized to apply these new techniques. The *corps du Génie* (Army engineering corps) for fortification engineers and the *corps de l'Artillerie* (Artillery corps) were created by the French army in 1691 and 1755, respectively. On the other hand, several civilian corps of state engineers were set up such that the corps of *Ponts et Chaussés* (Bridges and Roads) appeared in 1716 and the *corps des Mines* (Mining corps) at the end of the eighteenth century. These military and civilian corps gave rise to engineering schools. The correspondent two schools *Ecole des Ponts et Chaussés* and *Ecole du Génie* were the most important engineering institutions before the Revolution.

At the beginning of the French Revolution, France fought against the Europe. The engineering schools that appeared under the *ancient régime* were in crisis and their subsequent depopulation can be associated with the state of suspicion of any corps or other form of organization contrary to the revolutionary ideal. However, the engineering corps was maintained in order to defend the new Republic.

The creation of the Polytechnic School was Monge‘s project, created for the purpose of preparing engineers for military and civilian careers with different requirements dependent upon the respective engineering corps. In 1794, the *École Central des Travaux Publics* (Central School of Public Works) was created, with the name being changed to *École Polytechnique* (Polytechic School) in the following year. The new school put a very ambitious challenge to itself, namely the preparation of engineers to build fortifications, bridges and roads for exploring and operating mining resources, along with the training of shipbuilding engineers, geographic engineers and all kinds of other engineers to work in all technical domains. In addition, the projected time for this preparation was three years. The strategy for accomplishing this task was a universal scientific knowledge of tools and methods that could be applied over a large range of practical problems.

Among the staff of the new engineering school were the mathematicians Lagrange (1736–1813) and Laplace (1749–1827) and the chemist Bertholet (1748–1822). Other famous scientists took part in the teaching of mathematics and mechanics, such as Monge and Prony (1755–1839). Soon afterwards, Fourier (1768–1830) and Poisson (1781–1840) joined them.

Fig. 5.3 École Polytechnique (5, Descartes Street)

Within a few years, the Polytechnic School became a great center for the creation and dissemination of knowledge in Europe. Paris also became the undisputed capital of scientific thought in Europe and, at the beginning of the nineteenth century, the scientific spirit was firmly established in France.

A reformation of engineering instruction was necessary because war with other European countries had stimulated the construction of fortifications, roads and bridges, and the development of artillery. This new context propelled France to apply scientific principles to industry, with the result that the new engineering school had to provide universal scientific knowledge as well as tools and methods applicable in a diverse range of practical situations (Fig. 5.3).

Other schools appeared afterwards, such as the *École d'Application de l'Artillerie et du Génie* in Metz, the largest of the pre-World War I applied schools, as well as other much smaller schools. The *École Polytechnique* thus became a central component of a more complex training structure which many historians refer as the *polytechnique system.*

The change in the institutional framework providing better conditions for scientific research gave rise in the eighteenth century to very important scientific developments. In this context, some fundamental works of scientific thought appeared. It is important to mention the following: in 1788, as we mentioned before, Lagrange's *Analytical Mechanics*, in 1789, Lavoisier's *Chemical Treatise*, in 1796, Laplace's *Exposition of the System of World,* in 1799, Monge's *Treatise of Descriptive Geometry*, creating a new mathematical branch. Jean Baptiste Lamark (1744–1829), in the period from 1794 to 1800, conceived of species' evolution and published his *Zoological Philosophy* in 1809. Obviously these works could only be possible through a significant change in scientific thought, but we did not mention other fundamental works, such as Coulomb and Lazare Carnot's investigations into applied mechanics, because we will study these contributions later in this chapter.[8]

The role of the eighteenth century in the history of scientific thought should be understood not only in regards to one discovery or one revolutionary theory, as occurred in the seventeenth century with the publication of *Principia*, but for the great quantity of scientific investigations and the spreading of interest in new fields of knowledge. It is also noticeable in *Light Century*, the consolidation of certain branches of nature sciences such as electrology, chemistry and biology. In other words, electrology received an important contribution from Benjamin Franklin (1706–1790), who demonstrated the identity of such phenomena as apparently different as electrical sparks and electrical discharges. The most eminent engineers and physicists to study this in field of investigation were Charles Coulomb (1736–1806) and the Italians Luigi Galvani (1737–1798) and Alessandro Volta (1745–1827), the latter discovering the electrical pile in 1800. This fundamental technological advancement, which allows for the storing of electricity, provided future developments in electrodynamics as well as the discovery of unknown substances by using the process of electrolysis. It also contributed to the achievement of the principle of energy conservation, the first law of thermodynamics.

Furthermore, we have, in chemistry, the revolution created by Lavoisier. With his work begins the age of modern chemistry, the fall of flogistic theory being a fundamental step towards chemical progress. Unfortunately, Lavoisier was executed by the Revolution which is attributed, among other reasons, to his position during the *ancient régime* as a kind of tax collector, which was thus used incorrectly to identify him with the old regime, in spite of his revolutionary work.[9] In biology, we can look at the contributions of the Swedish scientist Charles de Lineu (1707–1778), the French George Louis de Buffon (1707–1788) and the Italian Lazzaro Spallazani (1729–1799). Lineu and Buffon presented a development of great interest through the observation of phenomena and through science unification, along with a strong Newtonian influence. Buffon believed that life's essence is a physical property of matter.[10]

---

[8] In the mechanical field, as we mentioned before, the works which we need to note are those of d'Alembert, Euler, Coulomb, Carnot and Lagrange.

[9] Obviously, the political process led the great scientist to death and involved more complex reasons than a simple identification of him with the "ancient régime" due the place he occupied within it. Kahane (1974) in Chap. 2, did an analysis of Lavoisier´s social position, his political alignments and contradictions in order to try to explain his condemnation and death. He states: *Lavoisier, a bourgeois of origin, tried to become a finance man and therefore associated with the "Ancient Régime"; he also tried to become a big land farmer, being assimilated with the owners of agrarian richness, the rural aristocracy. He belonged then in a double sense to marginal fractions of bourgeoisie, that to be displaced from power by Revolution, and thus it happened with the incidents and the dramas of the process.*

[10] As we know, Buffon translated Newton´s *Calculus of fluxions*, to which we had previously referred in Newton (1994) in Chap. 2.

In the development of mechanics during the Revolutionary period, obviously the formalization of mechanics was the most important event, but it is worth noting Lazare Carnot and Coulomb′s works on applied mechanics as well as acknowledging the importance of certain mathematicians such as d′Alembert, Laplace, Euler, Lagrange and others to that formalization process.

In order to finalize our brief and general description of the scientific advancements in the eighteenth century, let us look at some specific developments in that period, mainly in the fields of mathematics and physics. The name of Jean-Antoine-Nicolas, Marquis of Condorcet (1743–1794), appears with important contributions in two fields of mathematics: integral calculus and calculus of probabilities. In the former, he developed a theory of integration by using infinite series. In the latter, his name appears as a pioneer in the application of calculus of probabilities to the social sciences. Using d'Alembert's studies, Condorcet discovered new meanings and new applications for this theory as well as amplifying its foundations. His contributions were made in the context of increasing interest during the last third of the eighteenth century in economic calculation for solving problems of road construction, communication and other industrial and commercial problems. Another important application for mathematics was the field of insurance. That period saw the rise of insurance for ships, insurance against fire, and life insurance owing to the growth of commerce and its specific demands. These new mathematical applications of the social sciences to economy and commerce show us the new possibilities for science at the beginning of the modern industrial society, as clearly announced by the French Revolution.[11]

It is important to take into account that the findings in the scientific field during the French enlightenment increased the confidence in the powers of reason and led the intellectuals of this period to believe that man can solve any problem by the available means of science, even economic and moral problems. As a result, many philosophical and political schools appeared in order to explain or propose solutions to mankind's problems.

5.3 The Mechanics of Lazare Carnot

Lazare Carnot can be considered an important link between d'Alembert and Lagrange. As we will see in this chapter, an important heritage left by d'Alembert was used systematically by Carnot. Therefore, it is of fundamental importance to establish the differences between d'Alembert and Lagrange.

d'Alembert's influence on Carnot's work is undeniable. This is quite clear if we look at their ideas about conceptual questions regarding mechanics. In his

[11] This is a panoramic vision of the main scientific developments in the century of the Revolution. For a more detailed study, we recommend the work of the group REHSEIS edited by Rashed (1988).

Dynamical Treatise in Chap. 2, d'Alembert asks: *What are the causes that produces or can modify bodies' motion?* He continues: *We know until now only two: some appear to us at the same time that its effect, or before, at the time that they also appear; they have its source in mutual and sensitive action of bodies, a consequence of their impenetrability : there are the impulsion and others derived of them; all other causes are only known by its effects, and we do not know its nature; such is the cause that produces heavy bodies fall in direction to the center of earth, the same that maintains the planets in their orbits,* etc.

Carnot adopted a conception of force into his mechanics similar to d′Alembert's stating that force is generated by elementary impulses. In other words, impulses were of fundamental importance to him. Even with respect to weight, he considered this force as resulting from elementary impulses. He states: *The weight and all forces of this nature act by means of insensible degrees and do not produce any sudden change. However it seems natural that we consider these forces as impressing, in infinitesimal intervals, shocks infinitely small to bodies which are moved by them.* On the other hand, it is possible to detect a Leibnizian heritage in the above quotation. When Leibniz defined a living force, we can find similarity with Carnot.[12]

With respect to Carnot, Lagrange's concept of force is the opposite. Thus, one reads: *We consider mainly acceleratory and retardant forces, whose action is continuous, as the gravity which tends to impress at each instant velocity infinitely small and equal to any particles of matter* Lagrange (1989) in Chap. 3.

We should note that d'Alembert and Carnot saw actions among bodies by means of shocks instead of continuous interactions, because these actions are realizable and become evident through their effects. Corroborating his vision, d'Alembert affirms: *All that we can see clearly in a body motion is that it displaced itself and spend an interval of time to do it... I, for instance, made the observation what is behind the causes, to visualize only the motion that they produce.* Yet, he remarks, in a well known quotation, the obscure and metaphysical character of forces, which is also a critique of the Newtonian concept of force.

In *Fundamentals Principles of Equilibrium and Motion*, Carnot explains his ideas about mechanics in a more general way: *There are two ways to see mechanics and his principles. The first one is by considering as a theory of forces, the causes that impress motions. The second is by considering as a theory of motions by themselves. In the first case we can establish the rationality about the causes in any situation, which impress or tend to impress motions to bodies. In the second case, we see the motion already impressed, acquired or belonging to bodies; and thus we search only what are the laws according to which these motions appear and propagates, being modified or being destroyed for each circumstance. Each one of these two ways to study mechanics has its advantages and disadvantages. The first way is generally considered the simplest; but it has the disadvantage of*

---

[12] This was already analyzed in detail when the conception of force was discussed in Chap. 2.

be founded on metaphysical notion and obscure that is what a force is. Therefore what clear idea can present to spirit in this subject the name of cause? There are so many types of cause! What can we understand in a precise language of mathematicians by one force, i.e. by a double cause or a triple of one other?... These causes are the willing or the physical constitution of a man or of an animal that by means of its action produces the motion? But what is a double willing or a triple one of another willing?

> If we sponsor the position of do not distinguish the cause from the effect, that means, if we understand by word force the quantity of motion even if it made arise in body in which it is applied, become intelligible, but then we go back to the second way to study the question, that is, mechanics really is a theory of communication of motions.

As presented above, the concept of force occupies a central position in Carnot's mechanics, as well as signifying a d'Alembertian heritage. This long quotation brings up several points. The first is that the Newtonian approach is not adopted by him, considering force to be an external action exerted upon the body under consideration. The second is that Carnot, in fact, adopts the second way mentioned above in order to overcome the difficulty of identifying the force. By doing this, Carnot adopts an analytical tradition which comes from d'Alembert and will also intersect with Lagrange and others. Therefore, Carnot should be considered a link in a chain that leads to Lagrangian mechanics. This statement involves a complex process, because Carnot's mechanics use extensive geometry and trigonometry in a sense contrary to Lagrange's elimination of any figure or sketch containing forces and mechanics now belonging to mathematical analysis, which is actually the Cartesian dream. As a consequence of the second method adopted by Carnot, mechanical laws become the laws of communication of motion, which was also his understanding of how a machine operates. His contribution to applied mechanics is a natural consequence of his visualization of a machine or a system for communicating motion. Carnot's theory of machines, as we will see, is the study of these forms for communication of motion.

The similarities between the mechanics of d'Alembert and Lazare Carnot are not restricted to the conceptual basis, as we have seen in the case of the concept of force. They also use a similar structure in terms of principles where mechanics are anchored. In his *Treatise on Dynamics*, d'Alembert chooses three principles as dynamic foundation: the principle of inertial force, the principle of motion composition and the principle of equilibrium. A correct understanding of this second principle is of fundamental importance to a deeper vision of Carnot's mechanics. This is also necessary for revisiting the d'Alembert principle (Oliveira 2004) in Chap. 3.

If we examine Carnot's *Essays*, which we will do in detail in this chapter, we see how remarkable it is that he uses only two principles to present his mechanics: the principle of equality between action and reaction and the principle of nullity of the relative motion subsequent to a shock (between hard bodies). Obviously, the first principle is different from Newton's because of the differences on conception

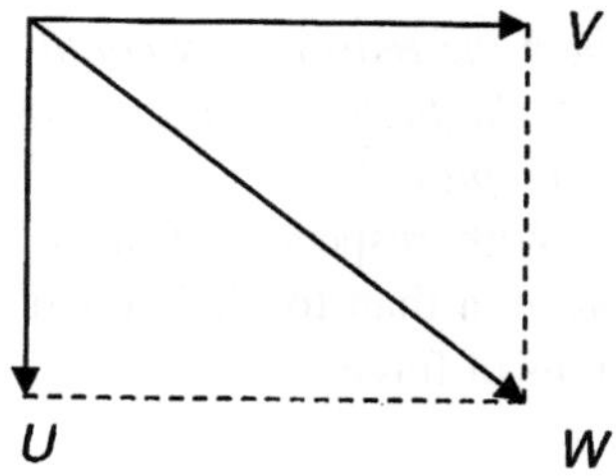

Fig. 5.4 Principle of motion composition

of force used by each of them.[13] This principle of Carnot's is much more than a law of equality between two quantities of motion. In the second principle presented, he uses motion decomposition as d'Alembert did in his famous principle. The two motions considered are impressed motion and attained motion, because the difference between them is exactly the force that a body would have if it didn't receive the impressed motion.

To understand the decomposition of velocities used by Carnot, the following diagram is shown:

$W = V + U$, where:

$W =$ INITIAL VELOCITY BEFORE THE SHOCK

$V =$ VELOCITY AFTER THE SHOCK

$U =$ LOST VELOCITY DURING THE SHOCK (Fig. 5.4)

Carnot's original contribution was his association of geometric motion with d'Alembert's decomposition of velocity. He defines geometrical motion as anyone compatible with the constraint of bodies. In the above diagram, if we consider a body shocking against a plane, if W is the body velocity before shocking the plane which coincides with direction of V, only it is reversible, because it maintains system constraints. U is not permitted, because the body would penetrate in the plane, which is by definition impossible, and consequently this motion is not reversible (geometric).[14]

At last, another fundamental characteristic of Carnot's vision about mechanics is the role of experience in our understanding of the physical world. He states: *It is, then from the experience that men built the first notions of mechanics. However, the fundamental laws of equilibrium and motion which are the basis of those notions appear on one hand naturally to reason and on the other clearly by the most common facts, which seem difficult to say if one precedes the other and if we have the perfect conviction about these laws, and if this conviction will happen without these causes. These facts seem to us very familiar to know that without*

[13] Indeed, the principle of action and reaction, known as Newton's third law, could represent a conservation principle if we consider the interacting bodies as a system. Thus, visualizing the third law in this manner is much more convenient for any study of the problems of shocks like Carnot's.

[14] The concept of geometric motion will be studied in detail in Chap. 6.

them the reason only could establish these definitions; if the reason is not used to link the facts by analogy, it appear to us very isolated in order to construct into principles.[15]

With respect to Carnot's general conception of mechanics, he is closer to Newton than to d′Alembert and the Cartesians in spite of their different conceptions of force.

5.4 The Work Concept in Carnot's Mechanics

In this section, we will try to follow the concept of work and its forms of transformation in the context of machines, now as a permanent acquisition of Carnot's mechanics.

As mentioned before, one of the most remarkable influences in Carnot's work was Bossut (1730–1814). In 1775, his *Treatise on Elementary Mechanics* appeared. This book is an excellent manual of mechanics, providing a general vision of this subject in Carnot's time when he was a student in Mézières, and was probably one of his initial and most important references regarding the study of machines.

Another mechanical study with respect to machines was the *New Hydraulic Architecture*, published in 1790, by Gaspard Richie de Prony (1755–1843), with the clear purpose of overcoming Bélidor's book, which, in fact, did not occur. The second volume of this work was dedicated to steam machines and has the importance of emphasizing one characteristic of this mechanical school of machines, which is to consider the motor, or more generically a drive force, as means to surpass a resistance, which points out the concept of work as the main concern.[16]

When some authors studied the concept of work in Carnot′s time, they always referred to Coulomb's memoir about friction, published in 1781, in which Coulomb won the prize and Carnot an honorific prize, as we mentioned before. However, some authors do not mention Carnot, such as in Prony′s course at the Polytechnic School. Otherwise, some important authors are referred to, such as Euler, Laplace and Lagrange. Bossut is cited by Prony in his hydrodynamics course (Gillispie and Youschkevitch 1979).

---

[15] It is worth remarking that Carnot made a rupture with the essentially rationalist tradition that came from d'Alembert and continued through Lagrange in which mechanics is seen as a branch of pure mathematics and experience as a kind of purely rational legitimation of that principle. Carnot would adopt a kind of empiricism that came from the English school and Newton as he continued to attribute to reason a fundamental role within the system of knowledge in establishing the causal nexus in the formation of laws.

[16] In Chap. 7, we will see in Coulomb's work one of the first and most consistent investigations into mechanical friction, one of the most important forms of resistance to motion. Later on, Coriolis and Poncelet studied one of the general forms of driven force utilization to surpass any resistance and created applied mechanics with other polytechnician engineers.

The first author to refer to Carnot's work appears in the first decades of the 19th century. He was André Guenyveau (1782–1861), who, in 1810, published the *Essay on Science of Machines*, a work different from Carnot's. He studied the equilibrium of machines and presented a series of practical applications, using the above-mentioned memoir by Coulomb in which the motion of machines can appear by means of shock or pressure. This distinction also indicates that continuous transmissions provide the best efficiency. But Guenyveau did not attribute to Carnot the origin of the principle used, in spite of the fact that in the preface of his *Essay*, Carnot is referred to as an author of a general treatise on machines (Gillispie and Youschkevitch 1979).

Jean-Nicolas-Pierre Hachette (1769–1834) is other author who cites Carnot's *Fundamental Principles of Equilibrium and Motion*, this reference appearing in the preface of his *Elementary Treatise of Machines*, published in 1811. Hachette remarks that Carnot, in the last chapter of *Fundamental Principles*, studies the whole theory of machines and the moving of applied forces. In addition, he mentions Carnot′s work as *the most profound savant and experienced engineer*. Paradoxally, Hachette notes few of Carnot′s achievements. In fact, the course that he taught was much more a collection of drawings of particular machines, also studying machine elements such as gears, pulleys, etc. (Gillispie and Youschkevitch 1979).

Alexis Petit (1791–1820) developed an investigation in the same sense that led to application of the work concept. In 1818, he published *About the Employment of the Living Forces Principle in the Calculation of the Effect of Machines* which is a very interesting work and coincides with our line of investigation. Petit was a physician who, together with Pierre Dulong (1785–1838), postulated the law that bears both of their names before his tragically premature death at 29. He presented general properties for motion as the conservation of living forces sponsored by him as the most efficient approach to machine calculation. According to him, it is the living force that permits in each particular case the best natural evaluation of the motor and the effect produced. The equation expressing the relation between these two quantities can provide the direct solution to the machine problem. In fact, what Petit was proposing was a method of energy balance for solving this problem (Gillispie and Youschkevitch 1979).

Petit assures us that scientists and engineers recognized that the living forces conservation principle is the most suitable for studying machines. Surprisingly, he affirms that *the theory of machines under this viewpoint is yet to be created.* This fact confirms that he didn't read Carnot's works, which had already adopted this notion that he praised (Gillispie and Youschkevitch 1979).

Petit guided his investigation, as did other researchers, assuming that the fundamental function of a motor is to surpass a resistance. If a machine is in equilibrium condition, it is sufficient to know the intensities of force, but if the machine is in motion, it is necessary to know the displacement of its points of application, which gives a central concern to the concept of work (Gillispie and Youschkevitch 1979).

Petit illustrates the applicability of the living force principle for calculating the effect produced by a machine using the equivalence in its dimensions of the quantities MgH and $\frac{1}{2} MV^2$. He also shows the classical example of bodies in fall. If the resistance is the weight with mass M elevated to a height H, the produced effect by the machine is found by multiplying the weight by height MgH. Or, as the acquired velocity of a heavy body that falls from a height H is such that $V^2 = 2GH$, the effect MgH is equal to $\frac{1}{2} MV^2$. Thus, for any type of resistance, the expression of the produced effect by the machine can always be expressed in the dimensions of the living force that is the same, through a product of a mass by the velocity squared. We should also consider that a motor contains some quantity of living force. What is true about the resistance to be overpassed is equally true for the motor: its capacity to surpass a resistance could always be reduced to a living force.

The approach used to solve the machine problem meant that calculations of any kind of machine were reduced to the determination of the relationship between the living force employed and the living force communicated to resistance. Then, it is possible to calculate under which conditions the machine efficiency is at a maximum. However, according to Petit : *the living force communicated to the resistance is equal to that pertaining to motor, subtracted by the lost living forces in sudden changes of velocity and that the motor conservatives after its action.* In other words, what Petit did is an energy balance representing the losses in terms of equivalence in living force, the same method followed by Carnot.

Since the beginning of the eighteenth century, it was common to calculate the power of a machine or a motor as a function of the height that they could elevate a given weight. Then, the quantities MgH and $\frac{1}{2} MV^2$ were practically equivalents in the operation of machines, and Petit, as well as his contemporaries, adopted these quantities as different forms of the same reality, i.e., the living force. They do not present any difference between work and kinetic energy, using their convertibility systematically.

Before denominating the product of a force by its displacement as work, but by its moment of activity, it is really in Lazare Carnot's theory of machines that we find the difference between living force and work. This achievement is what we will try to show in his theory. Sadi, his son, realized the meaning of the concept of convertibility.[17]

It was probably due to the re-edition by Navier (1786–1836) of Bélidor's *Hydraulic Architecture* that Carnot's general theory of machines finally became known by the new generation of engineers. This did indeed happen and some of them had a remarkable role in promoting and developing that theory until its transformations into a true science of machines. This subject will be studied in detail in Chap. 7.

The work by Navier referred to above was increased and improved with a series of comments and additions over time, but it is especially important to note that he summarized the history of the living forces principle from its origin. According to him, it was Galileo who first attributed certain definitions and concepts to the motion of machines. Galileo established that, for a given power in a given time, a

[17] We will study Lazare Carnot's concept of convertibility at the moment of analyzing the concept of geometric motion in more depth and try to relate this concept to the studies of Sadi Carnot on thermal machines.

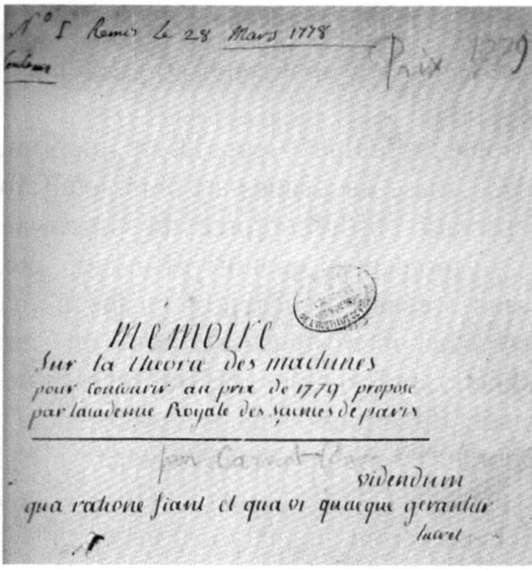

Fig. 5.5 Carnot's essay of 1779

combined effect is then produced which is equivalent and is measured by the product of the weight by the height to which it is elevated. However, according to Navier, the principle of living force conservation should be attributed to Huygens, being the result of a generalization made by him with respect to a system of bodies from a Galilean proposition according to which a body going down along any curve acquires the same velocity produced by a vertical fall from the same height.[18] In other words, if the friction is neglected, the mechanical energy is conserved as we know nowadays.

Finally, Navier made an evaluation of Carnot's contribution to machines, discussing the *Essay on Machines*, published in 1783, and the *Fundamental Principles of Equilibrium and Motion*, which appeared in 1803. Navier considers that Carnot creates a general theory of machines, thoroughly based in mechanical principles.

5.5 Lazare Carnot's Memoir of 1779

This memoir was submitted by Carnot to the Paris Royal Academy of Sciences in 1778 with the purpose of participating in a concourse promoted by the Academy. The essay is divided into two parts. Firstly, paragraphs 1 through 26 are dedicated to the description of friction experiences. The second part is divided into three sections. The first studies the principle of *machines in general*, from paragraphs 27 to 50. The second section, from paragraphs 51 to 79, is the study of seven classes of simple machines in equilibrium. The third section, from paragraphs 80 to 85, studies simple machines in motion (Fig. 5.5).

[18] We are referring to Galileo´s studies about the fall of bodies on an inclined plane which led him to his inertia law.

In this first memoir, Carnot made a review of general dynamical principles, visualizing their application to machines, which is done in the second memoir of 1781.

Some preliminary remarks are important before examining Carnot′s mechanics. Indeed, these mechanics are completely founded in general principles, mainly the conservation of living force, the conservation of quantity of motion, the conservation of moment of quantity of motion, etc. Regarding force, as we saw before, to Carnot, we can perceive it by means of quantity of motion.

Another interesting peculiarity of Carnot's mechanics is the extensive use of geometry and trigonometry in order to derive the conservation principles in convenient directions. This is a substantial difference between the analytical and algebraic Lagrange approach where all kinds of figures and any type of sketch are abolished. Still, Carnot's work can be considered an important contribution to mechanics from an analytical viewpoint with respect to the development of a theory of machines in the framework of Newtonian mechanics.

As we commented before, the application of conservation principles in Carnot's scheme is made by using d'Alembert's decomposition of velocities, such that the actual velocity of a body is ever decomposed into two components: the velocity that the body would have without any perturbation and the velocity which is destroyed by shocks through the motion.[19] When the concept of geometric motion is applied associated with this decomposition, we have an interesting and original application of the conservation principles. Indeed, this idea of geometric motion is a *sui generis* version of the principle of virtual works, with an advantage, that is, an attempt to generalize it.

The fundamental assumption in the development of an applied mechanics theory is that these two *essays* were preliminary studies to his masterpiece, the *Fundamental Principles*. As we are focusing our attention on a general theory of machines instead of a general analysis of mechanics, only the aspects directly associated to this theory are taken into account. Obviously the passages from the E*ssays* to *Fundamental Principles* will be studied from the evolutive viewpoint.

The problem to be solved by Carnot, which is the motion of any machine, is quite similar to the dynamics of a system of particles discussed by d'Alembert in his principle. In fact, it is a system of n bodies or particles, with the possibility of most of them being constrained or connected among them, and the system as a whole constrained by any manner, and to this system we impress any initial condition to one or more particles. The solution is to find the subsequent position for any particle of the system. If we call m the mass of any corpuscle or particle of the system, V the velocity that it would have if free, i.e. without reactions from the other parts of the system, u the velocity that it actually has and y the angle between V and u, we will have: $\sum mu(V \cos y - u) = 0$, that is, the summation of the products of quantities of motion for each corpuscle by the lost velocity, calculated in the direction of u is zero.

[19] See the d'Alembert principle in its original form in (d'Alembert 1921) in Chap. 2.

This can be proved decomposing V in two components, u and the other which is the lost velocity in shocks, so that the velocity estimated in u direction is $V \cos y - u$. This is the proposition which initiates the memoir part related to machines and enunciated in paragraph 27. In paragraph 29, Carnot postulates what he calls fundamental theorem, in the following terms:

If any system of hard bodies acting among them of any manner being immediately or by means of a machine, calling m any molecule of the system, V the velocity that it would have free from interactions in a given instant, u the velocity which it actually acquired due to reciprocal action of different parts of the system, y is the angle between the directions of velocities V and u; I affirm that we have: $\sum mu(V \cos y) = 0$.

Regarding that the decomposition V cos y is u itself, we have, in fact, $\sum mu^2 = 0$, which means the conservation of total kinetic energy of the system of particles. What we have to take into account is that Carnot doesn't differentiate between integral and summation symbols, as well as discrete and continuous systems. Also note that the above equation actually represents an energy balance, where the system communicates motion to another part of it such that the process is conservative.

In Corollary 13, a reference to work appears for the first time. In a complete quotation, one reads:

> As many parts of the machines are designed to elevate weights i will stay some moments about the case or in equation $\int mu\,du - \int mp$ ds $\cos x = 0$, found in corollary 3, where p is the gravity force and M the total mass of the system, H the height where go down the gravity center during the time t, V the velocity acquired from the height H, h the height from where the mass m has gone down in the same time; is, then clear that we have ds cos x = dh and $\int mp\,ds$ cos $x = \int mp\,dh = Mp\,dH$, and then 2 $\int \int mp\,ds$ cos $x = 2 \int Mp\,dH$ or $2MpH = MV^2$, then $\int mu^2 = \int mk^2 + MV^2$.

The penultimate equation obviously is an energy balance or, which is the same, the work done by the weight since H transformed itself into kinetic energy equals ½ MV^2, k being the initial velocity of the corpuscle, the last equation represents the modulus of any one of the three vectors as a function of the other two.

These are the first algebraic equations in Carnot's first memoir involving the quantities that later on will be known as kinetic energy and work and now are denominated by living force and moment of activity or moment of action. The last two expressions are the forms by which Carnot referred to work. In fact, this treatment given by Carnot is the application of conservation of the living force principle, as we have mentioned frequently. This question will be the main concern of our analysis in the first decades of the nineteenth century.

Following Carnot's memoir, he made many particular applications, including cases where the system is in an equilibrium situation, but always starting from motion situations. In item 48, he discusses the quantities which vary when the machine is in motion such that we can take advantages if we know this form of variation. Carnot affirms: *The advantage of machines is that we can vary the factors F, u, t, but the product must be ever MgH, since that it is a weight or a similar quantity that is involved, even being other kind of resistance.*

What Carnot explains many times in different ways is the conservation of energy within the field of mechanics. Thus, it is possible to vary in an appropriate form the three quantities of force, velocity and time, but the work or energy used in any process never overcomes the constant quantity *MgH*.

The second part of this memoir deals with the equilibrium of simple machines, and so we will move on to the second memoir, also called *Essay*, where a general theory of machines is examined.

5.6 Lazare Carnot's Memoir of 1781

This second memoir was submitted by Carnot to the Paris Royal Academy of Sciences for the concourse of 1781. It contains 191 paragraphs on 107 pages. It is divided into two parts. The first, from paragraph 1–100, consists of an experimental study of friction. The second part studies the same subject investigated in the first memoir, but in a more developed manner, also divided into two parts. Firstly, from paragraphs 101–160, it studies *Machines in General*. Secondly, from paragraphs 161–191, an *application of the experiences in simple machines* is described (Fig. 5.6).

At the beginning of Carnot′s investigation, he emphasizes a kind of theory which would be a general one and common to any machine, as follows:

> Each machine can present particular properties. Here we will study only that which are common to anyone; in order to examine the machines under a general viewpoint we will avoid repetitions that we need to do when we deal that properties which are common to anyone; we use the advantage of reducing the principles at least number possible but permitting realize these truths so disconnected at the first glance.
>
> In spite of differences among all machines, their properties are in somehow understood in one simple law from which we can derive easily all that refers to a particular one. This law will be explained and demonstrated rigorously after to speak a little bit about one general and simple idea within this subject.

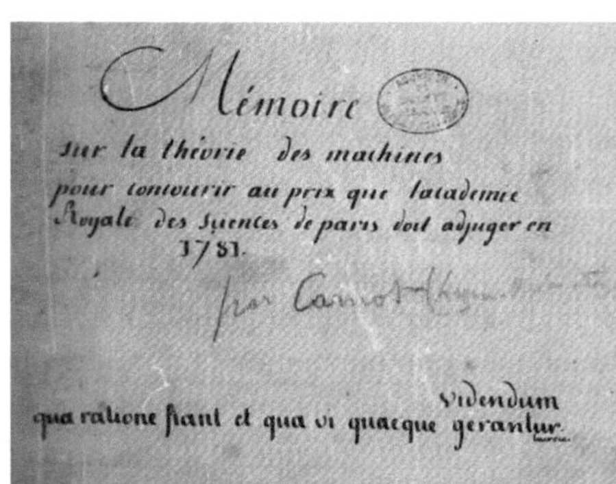

Fig. 5.6 Carnot's essay of 1781

Carnot extensively describes the question of machines in equilibrium in different situations, the simplest one appearing where only the weights and its constitutive parts are taken into account with the position of center of mass obviously at the lowest position. In paragraph 109, Carnot come back to the main concern, that is, the general theory of machines, as follows: *A general machines science in spite of all mechanics can be reduced to the following question. Knowing the virtual motion of a system of bodies, i.e. that motion which each one body acquires if it would be free to find the real motion that it will have in the following instant due to reciprocal action among the bodies assuming that each one of them has inertia as any matter. However, this problem is much simpler than among bodies in which some of them have no inertia and, of course we couldn't have a general theory of machines before to solve this problem completely: we will try now to solve this problem.*

It is important to note two things after reading the above citation. Firstly, a science of machines or a general theory of machines as denominated by Carnot belongs perfectly to the conceptual mechanics framework. According to him, the problem of a science of machines is a problem of general mechanics. Secondly, if we are dealing with a theory, it is necessary to create a simplifying hypothesis, to create models as is normally done in science. With respect to its modeling, Carnot supposes that the question of inertia is a problem to be solved previously. In terms of mechanical modeling, he states that, if there is interaction among the bodies, and all of them are perfectly elastic, it is possible to consider all of them hard, but linked to springs conveniently built in order to reproduce those elastic properties.

In paragraph 111, he makes an important statement: *The unique principle which can lead to the solution of the problem under consideration is the following: The reaction is always equal and contrary to the action. Is from this simple and undeniable law that we start, universal law which submit equally all and any body, being during the shock, the pressure or the attraction itself and all known nature phenomena, but it is just to know its effect on the shock and on the pressure.*

We should take into account that the principle of action and reaction as used by Carnot is the same as the conservation of the quantity of motion and not an equality between forces in a Newtonian sense. Also using the previous discussion about modeling, we consider any two corpuscles isolated from the system which are separated by small uncompressible bars with motion communicated among neighboring bodies until they spread out over the whole system. Then we have:

M' and M'' = mass of these corpuscles

V' and V'' = velocities that they have at the subsequent instant

F' = action of M'' over M', i.e. the force or quantity of motion that the first impresses upon the second

F'' = the reaction of M' over M''

q' and q'' = the angles between the directions of V' and F' and by the directions of V'' and F'', respectively.

The relative velocities between M' and M'' will be $V'\cos q' + V''\cos q''$ and because the bodies will go together, one reads:

$V'\cos q' + V''\cos q'' = 0$. Then, by the principle that the reaction equals the action in the contrary sense, one reads: $F'V'\cos q' + F'''\cos q''' = 0$. If one of them

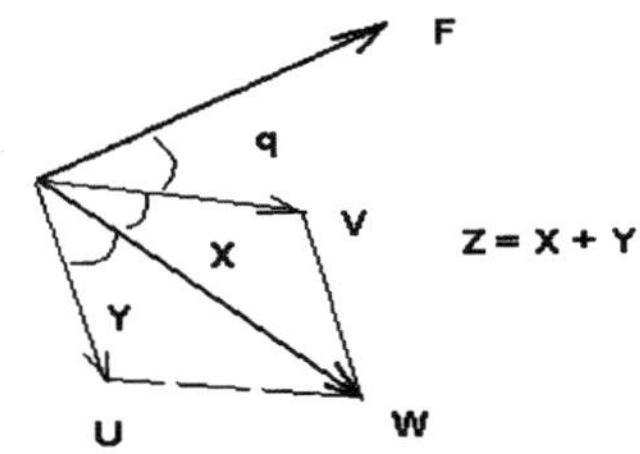

Fig. 5.7 Decomposition of forces and velocities

is fixed or an obstacle, one reads: V' cos q'= *0* ou V' cos q'= *0*. To the wholesystem and the corpuscles being taken two by two, one reads:

$$\int F'V'\text{cos}q' + \int F''V''\text{cos}q'' = 0.$$

Reestablishing the previous problem and creating a more d'Alembertian version, as Carnot normally did, one reads:

M = mass of each corpuscle of the system
W = its virtual velocity
V = its real velocity
U = the lost velocity such that W is the resultant of V and that velocity
F = force that impresses upon M each one of the adjacent corpuscles and by means of them it obviously receives all motion transmitted to different parts of the system
X = angle between the directions W and V
y = angle between the directions W and U
Z = angle between the directions V and U
q = angle between the directions V and P (Fig. 5.7)

For the whole system, we have $\int Fv\cos q = 0$ or $\int VF\cos q = 0$. Because the velocity before the shock is W, this estimated velocity in the direction of V will be W cos X, then V−W cos X is the velocity acquired by M in the direction of V then $M\,(V - W\cos X)$ is the addition of the forces F which act on M each one estimated in the direction of V, then $M\,(V - -W\cos X)$ is the same addition multiplied by V or for each corpuscle responds one peer, such that the sum total of all these additions is $\int VF\cos q$, then $\int MV\,(V{-}W\cos X) = \int VF\cos q$, i.e., $\int MV\,(V{-}W\cos X) = 0.\{\}W$ is the resultant of V and U, then $W\cos X = V + U\cos Z$.

Substituting this value in the previous equation, we have: $\int MVU\cos Z = 0$, which is the equation called fundamental, referred to in Carnot's first memoir.

> *At this stage of the memoir, Carnot introduces the concept of geometric motion, as follows:If a system of bodies starts from a given position in an arbitrary motion, but such that it was also possible to submit the system to another equal motion but opposite, each one of these motions will be denominated by geometric motion.*

Carnot gives a series of examples of geometric motion, emphasizing that its main characteristic is the reversibility, that is, if the system admits an equal and opposite

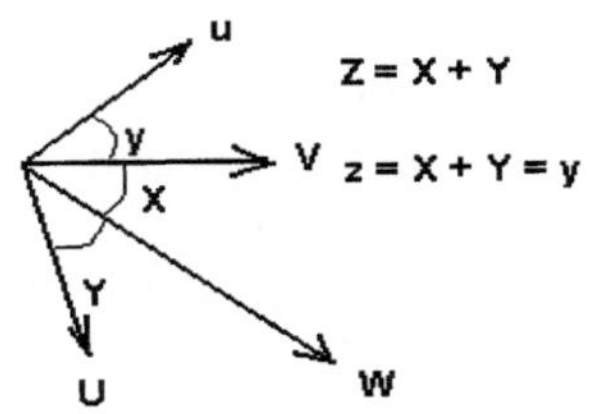

Fig. 5.8 Fundamental equation

$$\int muv \cos z = 0 \quad (1)$$

$$\int muv \cos z = \int muv \operatorname{sen} y = 0$$

motion, as possible. In this case, motion depends only on geometrical configurations and not on dynamical rules.[20]

In paragraph 116, Carnot proposes solving the following problem:

If the virtual motion of any given system of hard bodies is known (acquired motion if each one of the bodies were free), how do we subsequently find the real motion?

Solution:

Define the quantities as follows:

m = mass of each corpuscle of the system;

W = given virtual velocity;

V = searched real velocity;

U = lost velocity of each corpuscle in shock such that *W is the resultant of V and this velocity*;

u = the geometric motion.

Suppose that we give to the system an arbitrary geometric motion such that u is its geometric velocity.

X = angle between W and V; x = angle between W and u;

Y = angle between W and U; y = angle between V and u;

Z = angle between V and U; z = angle between U and u (Fig. 5.8).

Carnot presents several forms in which equation (1) could be written; all of them are related to the scalar product of the involved vectors.

By considering the previous vectors and their decomposition in the three coordinates axes as Carnot indicated:

---

[20] In Antonio Drago´s paper entitled: *Le lien entre mathematique et physique dans la mécanique de Lazare Carnot,* on (Charnay 1990), p. 501, we find several important remarks about Carnot's notion of geometric motion. The author emphasizes that, to understand this notion well, it is necessary to eliminate certain ambiguities left by Carnot, the obscure points being: (a) Carnot gives two different definitions of geometric motions. (b) He states several times that geometric motions are infinitesimal, and yet, at the same time, he considers them to be finite. (c) Carnot gives local definitions of geometric motions, but later on, he uses this concept from a global viewpoint, i.e., to the whole system. (d) Carnot uses geometric motion in an equation, called the second fundamental equation, where a logical derivation is neither clear nor incorrect. The author only considers geometric motion as a global motion and infinitesimally assumed or accomplished over the system if the opposite motion is possible. In other words, the fundamental characteristic of the geometric motion is its reversibility.

W — components $W'W''W'''$ $Uu\cos Z = U'u' + U''u'' + U'''u'''$
V — components $V'V''V'''$ $\int mU'u' + \int mU''u'' + \int mU'''u''' = 0$
U — components $U'U''U'''$ $\int mu'W' + \int mu'W'' + \int mu'''W''' = 0$
u — components $u'u''u'''$ $\int mu'V' + \int mu'V'' + \int mu'''V''' = 0$
The equations above are used to find the shock law for the particular cases.
As $W^2 = U^2 + V^2$; $\int mW^2 = \int mV^2 + \int mU^2$, he concluded:

> During shocks of hard bodies, immediately or by means of any machine without spring, considering or not the mass of the machine, the addition of living forces before the shock is equal to the addition of living forces after the shock plus the addition of living forces which would appear if the velocity of each part in motion was equal to that which it was lost due to shock.[21]

Carnot then applies an energy balance similar to the case in which motion appears through very small variations, which he calls insensible. From the mathematical viewpoint, this implies making U an infinitesimal of second order and thus:

$$\int mW^2 = \int mV^2$$

After showing some mathematical identities which came from these kinetic energy balances, Carnot introduced certain definitions in order to amplify this area of his analysis.

Definitions:

(a) The difference between driven and resistance forces depends upon the angle between the force and the velocity direction. If this angle is less than 90 degrees, we have the first case, otherwise the force is resistant.
(b) If an absolute driven force P is moving with velocity u, the quantity $Pudt$ will be called the spent quantity of action during dt by this force, i.e. the spent quantity of action in an infinitesimal time by a driven absolute force is the product of this force by the trajectory displaced of the point of application in this infinitesimal time.

Then, given this same quantity in terms of a given time, if we call ds the displacement during dt, we have $\int Pds$. If P is constant as the weight of a body, we have Ps (Fig. 5.9).
u = Velocity
P = Absolute force
z = Angle between P and u
Pudt cos z will be called the quantity of action produced by force P during dt, that is, during an infinitesimal time. In a finite path, we have $\int P \cos zudt$.

It is easy to see that these definitions are, in fact, definitions of work, with the denomination of quantity of action. With these definitions Carnot develops his studies on equilibrium and motion of machines.

[21] It is a living force balance in which the losses due to shocks are transformed into equivalent terms to living forces.

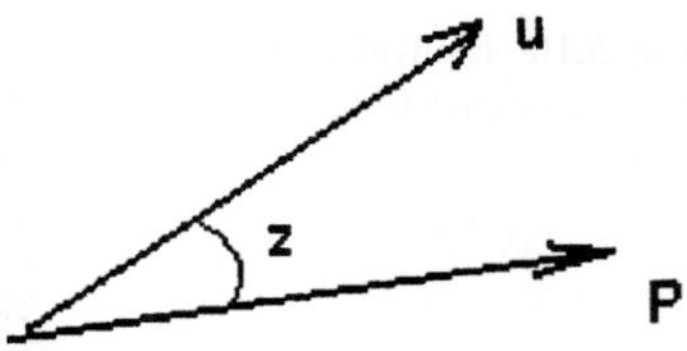

Fig. 5.9 Characterization of driven or resistant forces

Carnot began to study machines using a series of theorems useful for understanding his classificatory scheme for machines.

Theorem 1 General principle of machines in equilibrium

If a machine is in equilibrium and we impress an arbitrary geometric motion without any alteration in terms of applied forces, the quantity of action produced at the first instant by soliciting forces will be equal to the quantity of action produced in the same infinitesimal short time by resistant forces.

The above theorem is an application of the virtual velocities principle for simple idealized systems, that is, in the case of absence of friction, hard bodies and massless connections. Obviously, the constancy of force guarantees the equilibrium. Carnot then goes to the dynamical problem.

Theorem 2 General principle of machines in motion

If we impress suddenly motion to a machine with another motion being any geometric one and if we release the machine to itself, the conservation of living forces will happen immediately and at each elapsed instant of motion for any alteration of the driven forces.

The following nomenclature will be used:

$m =$ mass of each corpuscle
$V =$ actual velocity
$P =$ driven force
$U =$ velocity after the variation of actual motion in another geometric one
$X =$ angle between V and p
$\mathrm{Y} =$ angle between V and u
$\mathrm{Z} =$ angle between u and p
$\mathrm{p}' =$ driven force after the arbitrary variation
$z' =$ angle between p' and u (Fig. 5.10)

The virtual velocity of m estimated in u direction is $V \cos y + d\,(V \cos y)$ and the lost velocity by m during dt estimated in u direction is $p\,dt \cos z - dt\,(V \cos y)$.

With respect to Theorem 2, it is a kind of application of the living forces conservation principle having shown in the equation the figure for the living force on

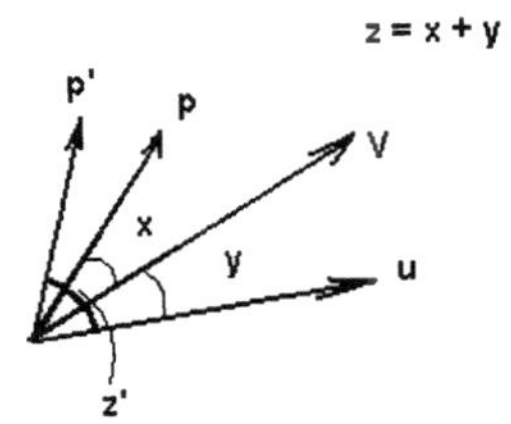

$\int mu\, du = \int mu\, dt\, p' \cos z'$

Fig. 5.10 Equivalence between living forces and work

one side and work on the other. Geometric motion, also cited in the theorem, is related to the invariability of system constraints and is not referred to as an application of the virtual works principle for a dynamical system.

The two theorems above are the fundamentals of machines study. Other cases are studied in the form of corollaries, and thus in particular cases.

Corollary 1 About weighted machines in motion

If we change suddenly the actual motion of a weighted machine in any other geometric motion and if we leave the system to its system of forces, the addition of living forces in each instant of subsequent motion is equal to addition of initial living forces (immediately after the variation) plus the addition of living forces which would have if each of the system would have its velocity equals to that due to height from where the center of gravity has gone down after the variation.

M = total mass of the system

H = height from where the center of gravity has gone down after the change W of the velocity due to height H

h = height from where each corpuscle m has gone down after the same instant

k = initial velocity and V the velocity in time t

z = angle between V and the vertical

We can obtain the following relations:

$\int mv^2 = \int mk^2 + MW^2$, which will be evident if the differential $\int mV\, dV = \int MWdW$ is an exact equation, but $W^2 = 2gH$, and then we have to prove that $\int mV\, dV = Mg\, dH$ or $\int \mathrm{mV\ dV} = \int mgdh$ or $mV\, dV = \int mgVdt \cos z$.

Corollary 2 About the machines in uniform motion

Always that a machine is moving uniformly (in which each point has the same constant velocity) the quantity of action produced in a given time by driven forces is equal to the quantity of action produced in the same time by the resistant forces.

One obtains the following relations:

$$mpVdt \cos x = \int mV\, dV \quad \text{but} \quad dv = 0 \quad \text{then} \int mpVdt \cos x = 0.$$

Corollary 3 About the weighted machines in uniform motion

Always that a machine without other applied forces but only its weights is moving uniformly, the center of gravity of the system stays invariably at the same height without going up or down.

The following relations are obtained:
$V = k$ then $MW^2 = 0$ then $W = 0$.

Corollary 4 About the simple machines submitted to periodical return

In a machine submitted to periodical returns the quantity of action produced during each period by driven forces is equal to the quantity of action produced in the same time by resistant forces.

Corollary 5 About the machines with friction taken into account

All that we can said about machines in general should be also extended to that where friction is taken into account or other any resistances, looking at these resistances as applied forces to machines but whatever motion, off course, these forces must be computed among that we have called resistances, i.e., among that which act in the contrary sense of real velocities of its points of application. Thus, for example, if the bodies are not submitted to driven forces, off course the velocities will be always decreasing and therefore we will look for in vain a machine subjected to friction with the capacity to maintain perpetually its primitive motion without alteration: and in addition we have found by means of experience that friction increases when the relative bodies velocity decreases; and starting from the lost degrees of velocity at each instant they will be growing, such that the motion cannot be suppressed but weakening progressively and to be extended as the friction could be, for example proportional to velocity as believed by some famous physicists and motion indeed weakening and cannot be perpetual.

The same reasoning can be applied to weighted machines and we arrive at the same point, because each time that the center of gravity goes up in order to go down further, this will be always with less velocity and therefore it will go up always progressively less until that will be impossible to go up.

The above corollary was transcribed in full in order to show that Carnot's mechanics do not consider machines as idealized systems, i.e., without friction and applying the principle of virtual velocities to equilibrium conditions and the principle of conservation of living forces to dynamical cases. As the long citation proves, friction plays an important role in Carnot's considerations. And yet, it is important to highlight that he normally converts in work or in living force that part which is lost such that his analysis is restricted to the mechanical field. In other words, in spite of the concessions to Carnot that even work could be spent by friction, it can be converted into an amount of living force.

In the final part of Carnot's memoir, we find general considerations about machines. The concept of work returns to the center of concerns when he asks, what is the true objective of a machine? Trying to answer, he affirms: *We have already said that is the possibility to vary in any way and according to the exigency of circumstances, the terms of the quantity of action F V t or Q produced by driven forces. If the time is valuable and if the action should be produced in a very short time and besides this we have only a force that can impress low velocity, we can imagine a supplementary machine to get the necessary velocity to that force. If on the contrary, it is necessary to elevate a huge weight with a weak power, but with capacity to impress high velocity, we could imagine a machine with which an agent could compensate by means of its velocity the missing force. Therefore, if the power is not capable neither of a big effort nor of a high velocity we could yet with the help of a unique machine to overcome the resistance under consideration, but it will be impossible do not spend a lot of time and that is important to understand, i.e. through this known principle: in machines in motion what we lose in time or velocity we win in force.*

This discussion is essentially the same as that already considered in Chap. 3, when we introduced the ideas of work and energy. The quantity that stays approximately constant and limits the action of a given machine is the work that it is capable of undertaking. With respect to the enunciated principle in the last phrase of the above citation, this has been known, at least empirically, since the times of Hero of Alexandria.

After these considerations about the forms of the working of a given machine, Carnot discusses the question of the efficiency of machines, looking for the most advantageous way to apply power to machines in motion such that the machine can produce the biggest possible effect. He states: *The general condition is that Q be a maximum which originates two considerations, the first referring to directions in convenient directions to give to the forces, the other the manner of providing conveniently its capacity. The first point is solved according to the sense of its velocities. The second question can be presented naturally: an agent is susceptible of two possibilities, one is the force, the other is the velocity; in order to obtain of this agent the whole effect of his capacity, there is a relation and a position between its force and its velocity, relation that we can only know by means of experience. For example, supposing that a man working continuously during 8 h per day with a lever can make a force of 25 pounds with a velocity of 1 foot/s, but if we induce this man to be much more fast, there are necessity of retarding because the man cannot maintain this continuous work with 25 pounds during 8 h, such that the quantity of action F V t produced by him will decrease. If, on the contrary, by decreasing the velocity, the force will increase more with respect to F V t, such that F V t will decrease according to the experience in order to F V t be a maximum is necessary to maintain approximately the velocity measured in feet per second and really to work around 8 h per day.*[22]

At the end of the memoir, Carnot introduces a kind of energy balance, obviously restricted to the field of mechanics, for both cases where shocks appear or not. For the case of a machine which elevates a weight P to the height H, the moment of action along the height H will be given by driven forces and will be $PH + 1/2 \int mu^2$.

For the case where shocks are present, we call q the quantity of action produced without shock, Q the quantity of action actually produced by shocks and h the height where the body with weight P will be at the moment of shock, X the addition of living forces of the system immediately before the shock and Y addition of living forces immediately after the shock. Then, one has:

At the moment of the shock $Ph + X/2$,

immediately after the shock $P(H - h) - Y/2$,

the quantity of action to be produced $PH + (X - Y)/2$,

i.e., that $Q = PH + (X - Y)/2$, where X is always greater than Y and $Q > PH$ or $Q > q$.

Carnot concluded that, in any way that we consider the shock, there will necessarily exist a loss of quantity of action. Finally, we can say that for the maximum transmission of motion, we must avoid friction and situations of shock as well, as these losses are also always measured by means of work, not yet with this denomination but referred to by Carnot as quantity of action.

References

Charnay JP (1990) Lazare Carnot ou Le Savant-Citoyen. Presses de l'Université de Paris-Sorbonne, Paris

D'Alembert JL (1921) Traité de Dinamique, vol 2. Gauthier-Villars et Cie Éditions, Paris

Fridberg A (1978) Sadi Carnot, Physician et les "Carnot" dans l'Histoire. La pensée universelle, Paris

Gillispie CC, Youschkevitch AP (1979) Lazare Carnot Savant et sa Contribuition a la Théorie de l'Infini Mathematique. Librairie Philosophique J. Vrin, Paris

Mota CG (1989) The french revolution (in Portuguese). In: Atica SA (ed) Sao Paulo

Oliveira ARE (2004) The contribution of coulomb to applied mechanics, Proceedings HMM 2004 Symposium, Kluwer Academic Publishers, p 217–226

Rashed R (1988) Sciences a l'Epoque de la Revolution Française (Recherches Historiques). Librairie Scientifique et Technique Albert Blanchard, Paris

[22] It is important to note that the concern with the efficiency of machines is still present in Carnot and will be developed in a wide sense by the polytechnician engineers Navier, Coriolis and Poncelet. However, we should mention the influence of Coulomb's work about the force of men on the argumentation used by Carnot.

Chapter 6
The "Fundamental Principles of Equilibrium and Motion" of Lazare Carnot

The rational mechanics of Galileo, Descartes and Newton was not, then, directly applied to machines and is not surprising that parallelly it was maintained a "corpus" of experimental knowledge, more or less formalized, addressed to practical constructors... It will be necessary to wait until the end of the eighteenth century to that Lazare Carnot's sciences of machines could be formally integrated to rational mechanics.

(François Vatin—The Work: economy and physics, 1780–1830, PUF, Paris, 1993)

In this chapter we will use Carnot's masterpiece regarding machines: *Principes Fondamentaux de Equilibre et du Mouvement*, edited in Paris in 1803 by Deterville which was written in 262 pages. In this analysis, we will follow the same structure presented in the book, the preface and two parts as follows:

(1) Preliminary notions. Assumptions considered as general laws of equilibrium and motion. Consequences of these assumptions.
(2) Development of previously settled assumptions as nature laws. The pressure and its laws through algebraic formulas. General considerations about driven forces applied to machines.

The two memoirs called *Essays on the Machines* were discussed in the last chapter. Now we will study *Fundamental Principles*, trying to avoid repetitions and common points with *Essays*, unless these points can emphasize the evolutive aspects of Carnot's work between the writing of *Essays* and *Fundamental Principles*.

6.1 Preface

Initially, Carnot said that this new book was, in fact, the same study (contained in *Essays*) in which he attached some developments in order to lecture about them more easily. With these changes, the work became something new, at least in form, and therefore he felt it appropriate to publish it under another title.

A. R. E. Oliveira, *A History of the Work Concept*, History of Mechanism and Machine Science 24, DOI: 10.1007/978-94-007-7705-7_6,

Carnot also paid attention to remarkably few changes between his previous work and this one, emphasizing the introduction of the least action principle. With respect to this novelty, he states: *Maupertuis, as we know, gave the first idea of this principle, when the motion varies by insensible degrees, as well as by means of sudden changes. But he only realized vaguely this principle, thus concluding about the final causes, he doesn't make any difference between the two case mentioned. There is, a great interest in applying the least action principle to both cases, giving to its enunciate the necessary understanding and mathematical precision, being necessary to do propositions with nothing in common, or better, that there are two exact and different principles, where the vague Maupertuis's principle is replaced by one applied exclusively to the case where the motion varies by insensible degrees, and another exclusively applied to shocks of bodies occurred with sudden variations.*

And yet, according to Carnot, Euler separated the first case from the second, transforming the first into a rigorous proposition and applying this principle to a particular case, which is a body submitted to attraction law. Lagrange, through this new calculus invented by him, known as variational calculus, proved with elegance that Euler's proposition has a general character for a system of any bodies submitted to attraction law, exerted as a general function of distances. In the other case, that of shocks and sudden changes of motion, he believed that it was possible to make the vague aspects disappear and to transform it in the form of a principle to any case of a system of bodies acting over them by means of shocks or sudden changes, exactly as a machine would do. He then affirms: *In this new edition, I developed what I have said in the first work, by showing that this principle occurs to the bodies with several degrees of elasticity, as well as to hard bodies.*

Carnot, in a form very distinct from Euler and Lagrange, develops his mechanics to study machines. To analyze a machine, the continuous transformations and insensible changes of motion must be replaced by sudden and, in some cases, discontinuous changes of velocities or displacements. Carnot also tried to generalize the Maupertuis principle to give to his theory a high degree of generality.

The difference between his approach in *Fundamental Principles* and that in the two *Essays* is remarkable. Because of the treatment of problems involving percussion of bodies and sudden variations of positions and velocities, this new theory cannot be based on the virtual velocities principle. In this context, he introduces a new method based on a new concept of motion that he called geometric motion. According to Carnot: *As result a new kind of theory to be applied on a class of motions ... these geometric motions are that which acquire different parts of a system of bodies, without neither perturb themselves nor the other and consequently these motions do not depend of the action or reaction among the bodies, but only upon the conditions of their connections, and thus being determined only by geometry and not dependent of the rules of dynamics.*

The characteristic of an extensive use of geometry gives Carnot's theory a touch of originality. Its attempt to bring the dynamical problems to the kinematic field when motions only depend upon the constraints of the system is also an attempt to generalize the principle of virtual velocities. In other words, some special class of dynamical problems could be solved by generalizing this principle.

Another novelty of *Fundamental Principles*, with respect to the previous *Essays*, is that Carnot refers to the experimental basis which mechanics should necessarily maintain. The second part of *Fundamental Principles* begins at the point where he considers that mechanical science is moving out from its experimental basis to become entirely rational, i.e. where the principles seem, to him, sufficiently settled and confirmed by experience such that mechanics only need reasoning. At this point, science will be susceptible to applying analytical calculations.[1]

This vision of mechanical methodology is interesting because it represents a rupture with Cartesian thought about science. To them, mechanics is a purely rational science and its principles could be derived only through mental operations. Carnot inverts this viewpoint and sees the form and the manner in which the knowledge is obtained in the mechanical field. According to him, the starting point is a safe experimental basis, and in the second step the analytical methods can be applied. In fact, even if we look at the exponents of analytical methods such as Euler and Lagrange, Carnot's methodology is correct. Euler wrote of Newton's second law in his differential form and Lagrange formalized rational mechanics using the principle of virtual velocities and the principle of least action. The new theory proposed by Carnot follows his understanding about the development of science.

Carnot finalizes his comments in the preface of his book, saying that, similarly to previous work, he leaves the applications of the principles developed for machines until the end.

First part: Preliminary notions. Assumptions considered as general laws of equilibrium and motion. Consequences derived from these assumptions

In spite of his d'Alembertian heritage, Carnot attributes a fundamental role to experience in the process of acquiring knowledge about the physical world. At the beginning of this first part, Carnot comes back to this question in order to discuss it in a deeper way, and, in this context, quotes the philosopher John Locke. Carnot mentions the *Essay on the Human Understanding*, agreeing with Locke's statement that *all the ideas come from the senses*.[2] According to Carnot, it is as correct

[1] Again, the relationship between Carnot's mechanics and Newtonian mechanics appears in the context of Cartesian thought where mechanics was a pure rational science.

[2] In Locke's work (Locke 1973), we can find a discussion of this question in book II entitled: *The Ideas,* Chap. I, page 165. In the subtitle *All the ideas come from sensation or reflection*, one reads: "Suppose, then, that a mind is a blank paper, without characters, without ideas; how does it can be provided? From where come this vast stock, which activates the boundless man´s fantasy with a variety almost infinite? From where he apprehend all the reason materials and knowledge? I answer, in one word from the experience. All knowledge is founded in it, and from it derives fundamentally our knowledge itself. Employed in the sensible external objects as well as in the internal operations of our minds, that are realized by us and reflected, our observation provides our understanding with all thought's materials. From these two sources of knowledge arise all our ideas, or that we have".

for any science, even the most abstract, as it is for mathematics. He argues and explains his viewpoint by saying that *sciences are a series of established reasoning made about the quantity.* However, he also recognizes that the relationship between science and experience is different. Pure mathematics are less influenced by experience than other sciences; following an order of influence beginning with mathematics, we have mathematical-physics, and later physical sciences, etc.

Another important point, a criterion used in his book to give a coherent structure to this study, consists of identifying the point where each science broke with its experimental character to become completely rational. By using this criterion, it would be possible to reduce to a minimum number of truths what will be taken from the experience by any science, and once this truth has been settled sufficiently, we can combine through reason only that which could encompass all scientific ramifications. This is a modern discussion on the philosophy of science, which appears almost a full century later with Ernst Mach (1838–1936), the philosopher and scientist who postulated that the purpose of science was to give the most economical description of nature as possible, in order to provide conceptions which can help us better orient ourselves to our world. If science is uneconomical then it is useless in this regard. Thus, Mach maintained that economy must be a guiding principle in accepting or rejecting a theory.

The ideas of Carnot combining d'Alembert's rationalism with Locke's empiricism, with the emphasis on experience over the rational elements in the construction of our conceptual framework, approximates Carnot with Newton and English empiricism with respect to the epistemology of mechanics. Consequently, Carnot, in spite of the influences received from d'Alembert, is philosophically distant from the French rationalist school.

To follow the developments made by Carnot, it is necessary to present his definitions of the fundamental mechanical concepts: space, time, matter, equilibrium, motion, etc.

Definitions:

m = mass; e = space, or linear quantity; t = time;
all quantities of the form, or reducible to a form e/t are called velocity;
all quantities of the form $m\ e/t$ are called the quantity of motion;
all quantities of e/t^2 are called the accelerate force or retardant force;
all quantities $m\ e/t^2$ are called driven force;
all quantities of the form $m\ e/t$ or $m\ e/t^2$ are called the force or power;
all quantities of the form $m\ e^2/t^2$ are called the living force, or moment of driven force or moment of activity;
all quantities of the form $m\ e^2/t$ are called the moment of the quantity of motion or quantity of action;

Finalizing this set of definitions, Carnot defines equilibrium as the general destruction of all motion. This means that he considers equilibrium as a special case of motion, always visualizing motion in d'Alembert's decomposition scheme. The concept of work, obviously without this denomination, appears for the first time in *Fundamental Principles*, on page 36, item 57, as follows:

Let M a mass, P its weight, g gravity acceleration, dt infinitesimal element of time and H the height to which P was elevated. Following this new way of visualizing the forces, what must be employed to elevate P to the height H, will be PH; but H being an space displaced, can be expressed by the product of one velocity V and one time T; on the other hand, we have $P = gm = gdtM/dt$, and gdt is one velocity V'. Then $PH = MVV'T/dt$; then dt and T being two homogeneous quantities, PH will be the product of one mass by the product of two velocities, or by the mean velocity squared proportional between V and V'; then the force PH is a product of one mass by one velocity squared, as Mu^2, calling u the proportional mean between V and V'. Such description shows the natural origin of the notion of living forces. Another important question to be discussed is if the bodies' force in motion should be evaluated by the product of the mass by the velocity, or by the product of the mass by the velocity squared. This, as we have seen, can be reduced to a dispute of words.[3]

In paragraph 59, one reads: As we have seen the living force can be presented under the form Mu^2 of one mass by one velocity squared or under PH of one driven force by one line. In the first case it is indeed a living force; in the second case, we can give the particular denomination of a latent living force.[4]

Clearly, it is being stated that the two quantities are interchangeable, one could be converted into the other and vice versa. Carnot continued to use and develop these ideas about work, frequently in the context of its conversion into living force, although the denomination most frequent for the term work is the moment of activity. In the following citation one reads: *I will call of moment of activity spent by a driven force, to the product of this force by the path which is described by the point of application, estimated in the sense of this force; i.e., the product of this force by the way which describes the point where it is applied and by the cosine of the angle of projection, or the angle between the direction of this same force, and the direction of this same velocity.*

In a sub-item entitled: *Hypothesis that could be assumed as general laws of equilibrium and motion,* Carnot presents a series of hypotheses according to which the communication of motion will happen in a system of bodies. To Carnot, these hypotheses are the true laws of nature. We will enumerate all of them as follows:

(1st) Hypothesis—A body once put at rest does not leave by itself, and once in motion does not vary its velocity or the direction of this velocity.

(2nd) Hypothesis—If we impress new forces on different parts of any system of bodies in equilibrium, such that as a whole they are zero, the equilibrium will be maintained.

---

[3] Carnot refers to the dispute between Descartes and Leibniz about the quantity that is conserved in the universe, owing to the quantity of motion (according to Descartes) or the living force (according to Leibniz). If the main concern is what better represents the forces of bodies in motion, that would seem to be a semantic question, because each one can be representative, depending on the context.

[4] Work appears here as a form of living force.

(3rd) Hypothesis—If several forces, active and passive, are mutually in equilibrium, each one of these forces is always equal and in the opposite direction of the resultant of the others.

(4th) Hypothesis—The quantities of motion, or the driven forces which are destroyed in each instant in a system of bodies, can always be decomposed in other forces equal to two by two and directly opposed, following the straight line that links the bodies to which they belong; these forces can be regarded as destroyed, respectively, in each one of these bodies by the action of the other.

(5th) Hypothesis—The action of two neighboring bodies which act one over the other by shock, pressure or traction does not depend on its absolute velocity, but only by its relative velocity. In the case of two bodies that do not communicate directly but by means of intermediary bodies, the motion is transmitted from one to the neighbor, such that it is always solved in a series of actions which are exerted immediately between two neighboring bodies.

(6th) Hypothesis—The quantities of motion or the dead forces which are reciprocally impressed upon the bodies by means of strings or bars are guided in the sense of the strings or bars, and those which are impressed by shock or pressure are guided following the perpendicular from its common surface at the contact point.

(7th) Hypothesis—When the bodies that shock themselves are perfectly hard or perfectly soft, they always move after the shock accompanied by following the line of its reciprocal action which, following the previous hypothesis, is always perpendicular to its common surface at the contact point. When the bodies are perfectly elastic, they separate after the shock with a relative velocity equal to that which they have in the opposite sense immediately before the shock. When the bodies are neither perfectly hard, nor perfectly elastic, they separate with a relative velocity of approximate size, according to the degree of elasticity.

Carnot discusses these seven hypotheses, emphasizing their empirical character, as we have seen at the beginning of this chapter, when he describes how *Fundamental Principles* was constructed.

If we look at the seven hypotheses described above, it is easy to see that they encompass all kinds of motion of systems, especially the cases of transmission or communication of motion propagated through neighboring bodies, as well as problems of shock. Obviously, Carnot's objective was to develop a methodology to study machines.

The first hypothesis is clearly the law of equilibrium using another form of presentation. If a given body is stopped from changing this condition to acquire some motion, only an external action can modify both situations.

The second hypothesis still refers to equilibrium, but now with respect to a system of bodies. In this case, the concept of the resultant is introduced, i.e., the force which replaces the mechanical effect of all of them and, obviously, in a condition of equilibrium, it would be zero.

The third hypothesis consists of force composition, also involving the concept of the resultant, in which a vectorial operation is done.

The fourth hypothesis discusses the problem of shock between two bodies and the decomposition of quantities of motion involved, taking the direction where the shock occurs, and also using the principle of action and reaction. Action and reaction appear in different bodies.

The fifth hypothesis discusses the action between neighbouring bodies, and this action could be a shock, a contact pressure or a traction force through a flexible element. In the case of shock, what concerns us is the relative motion among the bodies. For the case of communication among bodies through another body, the action is transmitted by intermediary elements or by something else such that the transmission of motion is made by a chain of contributing bodies, but always by means of neighbouring bodies.

Obviously, it is possible to detect a subtle critique of the transmission of force at a distance as explained by Newton's gravitational attraction postulated in his *Principia.* To Carnot, as well as to Cartesians in general, in order to accomplish this action, a carrier agent is necessary, that is, a medium capable of transmitting this force throughout the space.

The seventh and final hypothesis discusses the shock among bodies that are perfectly elastic, inelastic, or with intermediary characteristics between these two extreme situations. In this analysis, for the two moments before and after the shock, Carnot introduces what is known nowadays as the coefficient of restitution, which is defined as the ratio between the relative velocity of separation and the relative velocity of approximation to the two bodies. For elastic bodies, this coefficient is equal to one, and for plastic shock, this coefficient is equal to zero. In the first case, there is no loss of energy, and in the second, the loss of energy is at a maximum with the two bodies going together after the shock.

After enunciating this set of hypotheses, which are entirely sufficient to define the basis of his mechanics, Carnot will try to test their validity by applying them to practical situations. His methodology is a kind of hypothetical-deductive approach. This happens in the following item: *Several consequences derived by previous hypotheses. What is called inertia force. Properties of forces which concurs to the same point. About parallel forces and the center of gravity.*

Carnot's idea about inertial force is the traditional one, having its origin in Galileo and Newton. Referring to a system of bodies, he defines this force as follows: *We call then inertia force of each one of the bodies in each instant, to the resistance which is opposed by it to its change of state, i.e., the reaction exerted by it over the system of bodies which modifies the system from a state of equilibrium to a motion; from motion to equilibrium, or from one motion to other: is, in one word, an equal and opposite force to that necessary to impress to this body, to change its actual state, to that where it will be in the following instant. Therefore, if we decompose the actual velocity before the shock into two other, where one is that which it will be acquired after the shock; the other multiplied by the body mass, will be that called its inertia force at shock moment.*

Carnot also advises avoiding the identification of inertial force with the lost quantity of motion. He adds: *The lost quantity of motion is the resultant of the quantity of motion produced by driven force and the quantity of motion produced by inertia.*

Carnot distinguishes inertia from inertial force: *The inertia is a property which cannot enter in calculation, but the inertia force is the true quantity susceptible of*

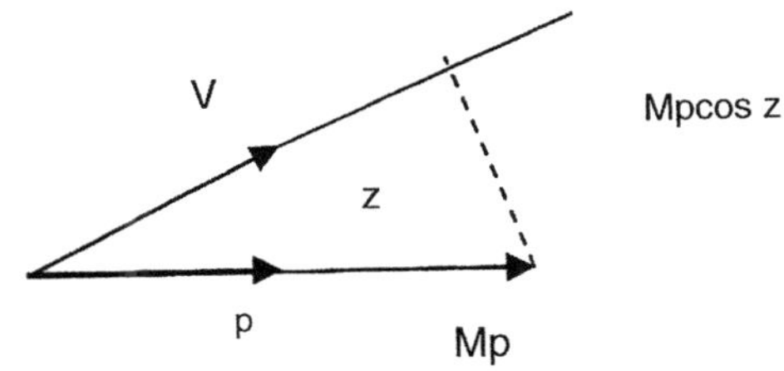

Fig. 6.1 Decomposition of quantity of motion

an exact evaluation. The inertia is the property inherent to each body which permits that it stays in his state of equilibrium or uniform and rectilinear motion; the inertia force is the quantity of motion which the body impresses to the other body whose state was modified by this body.

Then, Carnot mathematises some previous settled results. He calls *M* the body mass, *Mp* the driven force actuating on the body, *Mq* the inertial force, its effects *Mpdt* and *Mqdt* in the interval of time *dt*. The lost quantity of motion during *dt* will be the resultant of the two previous forces. If we call *V* the body velocity, its variation by *dV* in the interval *dt*, *z* is the angle between the velocity and the driven force. This force estimated in the *V* direction will be *Mp*cos*z* and the quantity of motion impressed in the sense of *V*, during the interval *dt* will be *Mp* (Fig. 6.1).

Defining *MV* as the quantity of motion, its variation in the interval *dt* will be *MdV* and then −*MdV* will be the effect of the inertial force, estimated in the sense of *V*; the resultant of these two forces is *Mp*cos*zdt—MdV*, in the sense that *V* will be the lost quantity of motion by *M*, in the interval *dt*. If this lost quantity of motion is the result of a pressure exerted by *M*, this force in each instant, along the direction of *V*, will be *Mp*cos*z—MdV/dt*.

If the body acquires a new velocity, given by *u* and if x is the angle between this new velocity and the force *Mp*, *y* is the angle between these two velocities *V* and *u*, *Mp*cos*x* will be the driven force in the sense of *u*; *V*cos*y* will be the component of the velocity *V* in the sense of *u* and, therefore, *d(V*cos*y)* the variation of this component in the sense of *u*. Then, −*Md(V*cos*y)/dt* will be the inertial force estimated in the sense of *u*. Now, the pressure exerted in each instant in the new sense will be *Mp*cos*x—Md(V*cos*y)/dt*.

Next, Carnot studies the equilibrium of a system of bodies for different systems of forces applied and includes the position of the center of gravity of the system. Without mentioning it, he enunciates the equivalent of Varignon's theorem in the following form:

For any system of forces in equilibrium around a given point, the addition of the moments of the forces with respect to any axe in space, is equal to zero, taking as positives the moment of the forces which tends to turn it in one sense and negatives all that tend to turn it in opposite sense.

Before defining the center of gravity of a system of bodies, Carnot studies the equilibrium of a parallel system of forces, applied to a system of bodies.

Before finalizing the first part of his book, Carnot discusses a new item entitled: *New consequences resulting from the previously settled hypotheses; accordance of these results with other facts generally recognized.*

In this item, Carnot will try to express the aspects of the theory already consolidated, by means of algebraic formulas. The following are the laws and principles considered by him to be fundamental to mechanics: the principle of virtual velocities, the principle of living forces conservation, the principle of the position of the center of gravity and the principle of least action. He proves the principle of virtual velocities solely through geometrical arguments, first using two forces and then generalizing to a system of any number of forces. He attributes this principle to Galileo. About Lagrange's contribution, he states:

Lagrange and his Analytical Mechanics, starts from the principle of virtual velocities between two forces only, as a fundamental truth also recognized, and he tries to extends, as we have done, but by a purely analytical way which he adopts in his beautiful work, this principle to any system of forces which act simultaneously.[5]

Similarly to Lagrange, Carnot also attributes great importance to the principle of virtual velocities as a founding principle of mechanics. Starting from this principle, Carnot tries to show that the virtual velocity of the center of mass of a machine in equilibrium is zero. Then, he enunciates Torricelli's principle in the following form: *For any machine with weights and in equilibrium, the center of gravity is in the lowest point possible.* Evangelista Torricelli (1608–1647) was a disciple of Galileo's who also enunciated in 1643 another principle regarding the speed of fluid flowing out of an open channel, which was later shown to be a particular aspect of Bernoulli's principle.

The form for representing the mutual equilibrium of several masses algebraically, of value M, V being its velocities, z the angle between the directions of the driven force Mp and V, requires one to use the principle of virtual velocities:

$$\int MpV \cos z = 0.$$

Also algebraically, the shock between hard bodies can be represented, if one considers the following quantities: let M be the mass of the body under consideration, W its velocity before the shock, V its velocity after the shock, U the lost velocity in shock and z the angle between V and U. Then, one reads:

$$\int MUV \cos z = 0 \text{ (For hard bodies)}.$$

Taking into account the vectorial composition (Fig. 6.2):

From the previous result, the last integral of the second hand side is zero. Then, one has:

$$\int MW^2 = \int MV^2 + \int MU^2.$$

[5] As we have seen, Poinsot (1975) in Chap. 3 makes the critique of the virtual velocities principle as a fundamental principle of mechanics.

Fig. 6.2 Conservation of living forces

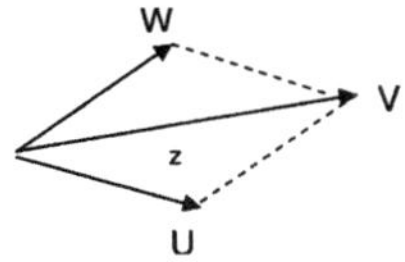

$$W^2 = V^2 + U^2 + 2VU \cos z$$

$$\int MW^2 = \int MV^2 + \int MU^2 + 2\int MVU \cos z$$

The above equation shows the conservation of living forces (kinetic energy) before and after the shock for hard bodies. And yet, with this result Carnot demonstrates Huygens principle, which involves shock of hard bodies with the particularity of infinitesimal motions, or, using the denomination of that time, for *insensible degrees of variation*. Carnot also remarks that U is infinitesimal, so that U^2 will be second-order infinitesimal, and thus, one has:

$$\int MW^2 = \int MV^2$$

This is the same as the conservation of pure kinetic energy, being established as the equality of these quantities before and after the shock. This last equation is also valid for elastic bodies, which is the same when there is no loss of living forces.[6]

Carnot concludes the first part of his work with a citation which is a kind of balance of the work accomplished up to this point:

These results of so different nature in appearance, and, therefore, all according with the proposed hypotheses should give us the justness of these hypotheses, a great confidence so that it is possible to wait for a science that is founded in part in experience. We are looking to future with these hypotheses, and our reasoning based on that, already confirmed, could provide the basis to the theory of equilibrium and motion.

Our conclusions from the above citation are that Carnot′s method is quite similar to the approaches of Galileo and Newton. Based on experience, a set of hypotheses is built which are submitted to validation through new experiences and new phenomena, and only after these steps is an attempt made to formalize this accumulated knowledge. In other words, Carnot was looking for a deductive-hypothetical scientific model.

Second part: Development of the previously settled hypotheses as nature laws. The impression of these laws by algebraic formulas. General considerations about driven forces applied to machines.

To define this part of Carnot's study we can use his words written in paragraph 135 as follows:

Now one tries to represent these principles by means of algebraic formulas. Yet, demonstrating the results rigorously only by reasoning, starting from previously established hypotheses and deriving the more general consequences. We will start by

[6] The decomposition of velocities using the d'Alembert principle transforms cosine law into a conservation principle of mechanics (principle of living forces).

shock of bodies, being immediate or operated by a machine. Later, we will derive, as particular case, the laws of motion to a system of bodies such that the motion variation is by insensible degrees. This theory will encompass all the fundamental principles of the communication of motions, and, consequently, of the mechanics.

Beginning with paragraph 136, Carnot introduces the concept of geometric motion which occupies a considerable space in his book. This concept is closely related to the principle of virtual velocities, consisting of an original contribution of Carnot′s important for our analysis. He defines geometric motion as follows:

All motion which is impressed to a system of bodies that does not vary its intensity of action exerted by them or that could be exerted by some bodies over others, if we impressed any other motions, will be denominated by geometric motion. He continues: *The velocity which acquires then each body will be denominated by geometric velocity.*

After defining geometric motion, Carnot illustrates this definition with some examples. The first is the case of two bodies approaching the point of shock; if we impress common motions upon it such that their relative velocities are not altered, these motions are geometric motions. The second example involves two bodies linked to each other through a string or bar. If we impress motions such that their relative velocities are not altered, the motions are also geometric. Other multiple examples are listed. Regarding the case of two bodies A and B at the extremities of a bar upon which we impress velocities to the extremities proportional to the lever arms and in opposite senses, it becomes a typical example of geometric motion. Finally, the case of two bodies A and B tighten the extremities of a string which passes through a pulley. If we impress the same velocity V upon the bodies A and B in opposite senses, we again have geometric motion.

Carnot notices that the definition of geometric motion can be generalized, because the concept is applied to any sort of body without distinction as elastic, soft, hard, solid and fluid bodies. The fundamental condition of its existence is that motions do not disturb any other, not appearing as any action over others. These motions are completely independent of the dynamics' rules, and depend only upon the constraints among the parts of the system, being determined only by the geometry of the system.

Following his study, we find a set of theorems and corollaries defining the properties of geometric motions visualizing its application later on. In fact, it is a construction of a theory for them. Let us examine the most important of these theorems and corollaries in order to analyze their content in the context of the principle of virtual velocities.[7]

Theorem I

When two bodies act one over the other by shock, pressure or traction, it is always or immediately, due to their neighborhood, or due a series of other

[7] Indeed Carnot is trying to generalize the principle of virtual velocities to systems in motion, in spite of being restricted to what he defines as geometric motion.

neighbor bodies among the firsts, and which transmit the action to the next by a series of immediate actions among these bodies neighbor intermediary.

Obviously, this theorem is a critique of the action at distance proposed by Newton in his universal theory of gravitation.

Theorem II

Since that a system of bodies acquires any geometric motion, and this system being perfectly free or perturbed by obstacles, it is always possible exert over it another coupled geometric motion equal to the first in opposite sense.

This theorem guarantees the reversibility of the geometric motions.

Theorem III

If two geometric motions are impressed to the same system of bodies and if this system is perfectly free or perturbed by obstacles, the resultant motion will be a geometric motion.

This theorem permits for the increase in complexity of these motions through a composition of simple components.

Theorem IV

In any system of hard bodies, if it is submitted by a shock or some instantaneous action, being immediate, by means of any machine without spring, the acquired motion after the shock will be necessarily a geometric motion.

Theorem V

Any geometric motion impressed to a system of any bodies, this motion is received by the system without alteration.

Theorem VI

In a system of hard bodies acting some over the others, being immediately, or by means of any machine without spring, if to the motion or to the shock that will happen, we decompose the general motion in two others, where one is that which occurs after the shock, the other will be necessarily that which will be destroyed, and these composed motions are such that the first will stays, it does not change, and if the second also maintains, there will be equilibrium in the whole system.

In spite of some difficulty in understand exactly what the theorem means, it expresses the d'Alembert principle precisely as it was originally enunciated. This principle was modified over time until it was associated with the principle of virtual velocities and later on with Newton's second law to obtain the equilibrium equations. However, the original d'Alembert principle is an attempt to solve the general dynamic problem of a system of particles which interacts by means of shocks or any kind of interactions. The problem to be solved is finding the final situation of all individual particles for a given initial condition in any number of particles. The principle states that the final velocity is the composition of two velocities. The first is the velocity which the particle would have if there was no perturbation by any other. The second velocity is the velocity that is destroyed by shocks during motion. The equations of equilibrium which are derived by d'Alembert principle came from the direction of this lost velocity, the direction of shock. A scalar product is taken along this direction, between the quantity of motion and this last velocity. This is the origin of the transformation of the d'Alembert principle in an equilibrium equation.[8]

Theorem VII

In a system of hard bodies acting among them, immediately, or by means of any machine without spring, if at the shock moment we decompose the general motion in two others, one which must be destroyed by the shock, and the other instead of it we substitute it by another any geometric, this motion will be that will appear after the shock.

Theorem VIII

The least sufficient force to break the equilibrium of a system, for any kind of system and any kind of motion acquired by it, is such that this force must produce a geometric motion; but to produce the same infinitely small, but not geometric, it is necessary a finite force.

[8] See Oliveira (2004) in Chap. 3.

Definition 1

When exists the equilibrium in a system of bodies, immediately, or by means of a machine without spring, and if we perturb this equilibrium by one infinitesimal force, the velocity acquired by each one of the bodies' system is called its virtual velocity and the general motion of the system is called virtual motion.

Theorem IX

All virtual motion in any system of bodies is necessarily geometric.

This set of nine theorems, some corollaries that we have omitted and one definition, established the bridge between geometric motions and the principle of virtual works. This study made by Carnot amplifies the importance of the concept of work within his mechanics, although the principle of virtual velocities also acquires a new meaning in Carnot's hands. His theory of geometric motions was in his vision of developing a science to occupy an intermediary position between pure geometry and the study of communication of motion, which is the mechanics of machines.

Theorem X

In shock of two hard bodies, if we consider that both can move, or that exists a fixed axe; the addition of the products of the quantity of lost motion by each one of them times its velocities after the shock, estimated in the direction of this quantity of motion is equal to zero.[9]

Theorem XI

In the hard bodies shock, with any number of bodies within the system, considering that some can move and others are fixed, the addition of the products of the lost quantity of motion by each one of these bodies, times its velocity after the shock, estimated in the direction of this quantity of lost motion is equal to zero (Fig. 6.3).

---

[9] It is possible to find the principles of conservation enunciated by Carnot using the definition of geometric motion.

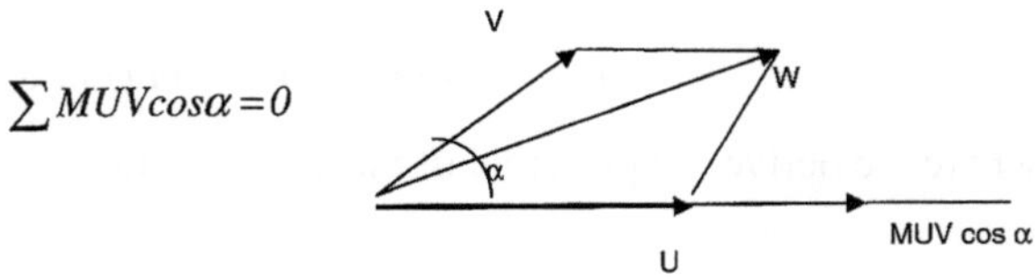

Fig. 6.3 Conservation of quantity of motion

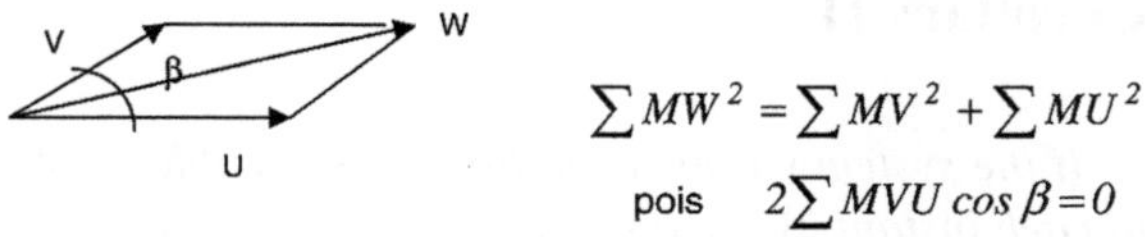

Fig. 6.4 Conservation of living forces

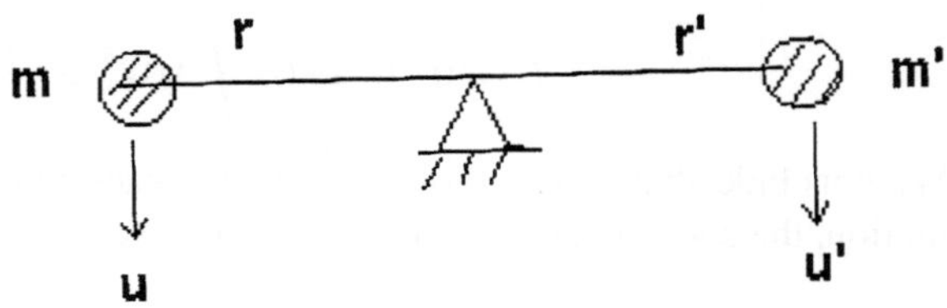

Fig. 6.5 Illustration of the principle of virtual velocities

Theorem XII

In hard bodies shock, for each number of bodies, and considering that the shock is immediate, or that it happens by means of any machine without spring, the addition of living forces before the shock, is always equal the addition of living forces after the shock, plus the addition of living forces which appear, if each one of the bodies would move freely only with the velocity which was lost in the shock (Fig. 6.4).

Corollary I

Since that several bodies are moving by the action of any forces but mutually in equilibrium, immediately, or by means of any machine without spring; the addition of the products of each mass by the velocity squared being this velocity that which it tends to move, is a minimum, i.e., minimum that not will be the addition of the products of each mass by the velocity squared which it would lose, if the system would acquire any geometric motion (Fig. 6.5).

$$Mg^2 + m'g^2 = m(g-u)^2 + m'\left(g+u'\right)^2$$

$$mu^2 + m'u'^2 = 2mgu + 2m'gu'.$$

Because u and u' are small

$$mgu - m'gu' = 0 \qquad mg/m'g = u'/u$$

where we derive the principle of virtual velocities.

Corollary II

If the system varies its motion by insensible degrees, the lost quantity of motion in each instant by the body will be infinitely small.

$$U = 0;\ U' = 0;\ \int MW^2 = \int MV^2$$

We conclude that, for a system of hard bodies with insensible variations of M motion, the addition of living forces is conserved.

Theorem XIII

In shock of bodies perfectly elastic, for any number of bodies, the addition of living forces before the shock, is always equal to the addition of living forces after the shock.

By definition, in elastic shocks, there is no loss of living forces. Using the equation previously established, one has:

$$\sum MW^2 = \sum MV^2 + \sum MU^2; \text{ being } U = 0,$$

$$\sum MW^2 = \sum MV^2.$$

Corollary IV

If the bodies are not perfectly elastic, but all of them with some degree of elasticity represented by n, i.e., such that the reciprocal action of the bodies instead of to be doubled, as it is happen in the case of perfectly elastic bodies, it was only this same force multiplied by n: it is clear that the direction of each one of the lost velocities would be yet the same, as well as the angle between W and U'''.

For the case of hard bodies, we have:

$$\int MUW \cos\left(W \wedge U\right) - \int MU^2 = 0 \tag{6.1}$$

If we suppose that with the degree of elasticity expressed by n, the lost velocity is U' and the remainder velocity is V', one has:

$$U' = nU;\ U = 1/nU'.$$

Multiplying (6.1) by n, one has:

$$n\int MU'W cos(W^\wedge U') = \int MU'^2 = 0,$$

$$W\cos\left(W^\wedge U'\right) = V'\cos\left(V^\wedge U'\right) + U',$$

$$\int MV'U'\cos\left(V^\wedge U'\right) + \int MU'^2 = 0$$

On the other hand, one has:

$$W^2 = V'^2 + U'^2 + 2V'U' cos\left(V^\wedge U'\right),$$

and thus:

$$1/2\int MW^2 = 1/2\int MV'^2 + 1/2\int MU'^2 + \int MV'U'\cos(V'^\wedge U')$$

$$\text{or} \int MW^2 = \int MV'^2 - (n-2)/nMU'^2.$$

For the case of perfect elasticity, $n = 2$, then

$$\int MW^2 = \int MV'^2.$$

For the case of hard bodies, $n = 1$, then

$$\int MW^2 = \int MW'^2 + \int MU'^2$$

It is possible then to put into a general formula the equation representing a balance of living forces for any degree of elasticity of the bodies involved in shocks.

Theorem XIV

In a system of any hard bodies in immediate contact among them, or belonging to any machine without spring, if a shock occurs, and if at moment of the shock, we decompose the motion which the system tends to move in two, one being that which must be destroyed, the other is such that, if it is suppressed alone, and we replace that by any other geometric motion, the addition of the products of the quantity of lost motion by each one of the bodies belonging to the system multiplied by the geometric velocity, estimated in the sense of this quantity of motion, it will be zero.

Theorem XV

Among all possible motions, a system of bodies perfectly hard acting some over the others by immediate shock, or by means of any machines without spring, such that occurs a sudden variation in the state′s system: all motions that will happen after the action will be geometric motions, such that the addition of the products of each one of the masses by the squared velocity which will lose by the system, is a minimum.

Let us consider:

M = mass of each one of the bodies of the system
W = its velocity before the shock
V = its velocity after the shock
U = its lost velocity during the shock
u = its geometrical velocity.

It is necessary to prove that $\int MU^2$ is a minimum since that M acquires a velocity V after the shock. This is equivalent to $\delta \int MU^2 = 0$. since, instead of V, we replace the geometric velocity u, which differs by an infinitesimal variation. Carnot presents a geometric proof of the proposition above using a series of decompositions and diagrams of convenient velocities. He already shows an alternative form of the minimization problem, by introducing the parameter time. If we suppose a given time t, each body M displaced the space X with a velocity U and thus one has: $U = X/t$. This formula can be written in the following form, since we divide by t, which is the same for all bodies:

$$\delta \int MUX^2 = 0.$$

In the long citation below, Carnot analyses his contribution in extending the principle of least action to the case of sudden motions. Due to its importance we transcribe the complete text from Carnot′s work. He states:

Maupertuis calls, in his Essay on Cosmology, quantity of action to the product of one mass by its velocity and by the path which it displaces. Thus, MUX is one quantity of action, and he advances in principle, that the quantity of action necessary to produce a variation in the body motion is always a minimum. This principle should be considered as the enunciate of the previous equation which is its algebraic translation. Maupertuis founds this principle on final causes; but as final causes are interpreted arbitrarily, we can in fact to say all that we need, that we never conclude precisely, if we do not support in mathematical demonstrations. Maupertuis proved that his principle is valid in direct shock of two free bodies perfectly hard, and for the case of two bodies perfectly elastic; although nothing was said, and his principle, even beautiful, was not supported by him, and by other geometers, with respect to sudden variations: at least I do not agree that somebody has demonstrated in a general form before the first edition of this work, where I established the equivalent principle given above; but only to hard bodies.

The demonstration I have finished now is more general, because it contains the bodies with different degrees of elasticity; although it also proves at the same time how are old the demonstrations based on final causes, because it shows that the principle is not so general; in fact it is restrict to the case where all bodies of the system have the same degree of elasticity. Although the theorem such as presented, seems to me simpler and easier to apply than that of least action where it is introduced useless the space displaced. But it is not less truth, that after these explanations, there is nothing vague in Maupertuis' *principle, and so it is rigorous and mathematically demonstrated.*

Following Carnot's exposition, he enunciates three blocks of propositions: principle of conservation of linear quantity of motion, angular quantity of motion and the percussion to any body in motion.

Theorem XVII

In a system of perfectly free bodies, being these bodies hard, soft or elastic, if a shock occurs:

(1) *The addition of the quantities of lost motion by every of these bodies, estimated in any sense after the shock, is equal to zero.*
(2) *The addition of the quantities of lost motion by one part of the system in a given sense is equal to the addition of the quantities of acquired motion, in the same time and in the same sense, by the other bodies of the system.*
(3) *The total quantity of motion of the system, estimated in any sense, stays the same as that before the shock.*

Theorem XVIII

In a system of perfectly free bodies, being these bodies of any nature, if a shock occurs:

(1) *The addition of the rotation moments of the quantities of lost motion by the whole system, with respect to any axe in space, and which tends to turn in the same sense of this axe is equal to zero.*
(2) *The addition of the rotation moments of the quantities of lost motion by one part of the system in a given sense around an axe is equal to the addition of the quantities of motion of acquired motions in the same time and in the same sense around this axe, by the other bodies of the system.*
(3) *The addition of rotation motions of the actual quantities of motion after the shock in a given sense around an axe, stays the same that before the shock.*

Theorem XIX

In shock of bodies, for the cases of hard bodies or not, with the action immediate or made by means of any machine without or with spring:

(1) *The addition of percussion moments of the whole system with respect to any geometric motion is equal to zero.*
(2) *The addition of the moments of activity of the whole system before the shock with respect to any geometric motion is equal to the addition of the moments of activity after the shock with respect to the same geometric motion.*

The concept of work re-enters Carnot′s text in the form of the relationship between work and living forces for the case of motion variations by insensible degrees.

Theorem XX

When a system of hard bodies, free or with constraints with respect to a machine without spring is moved by any driven forces, with insensible degrees of motion variations, if during any instant we call m each one of the masses of the system' corpuscles, V its velocity, P its driven force, u the acquired velocity if we suppress the actual motion, replacing it by other geometric motion; dt being the interval of time, we have the following equations:

$$\sum mVdV - \sum mVPdt\cos(V^\wedge P) = 0 \tag{6.2}$$

$$\sum mud\left[V\cos\left(u^\wedge V\right)\right] - \sum muPdt\cos\left(u^\wedge P\right) = 0. \tag{6.3}$$

The above equations link work with living forces. In Eq. (6.2), the virtual velocity is in the trajectory sense. The Eq. (6.3) prescribes the work related to displacement u of geometric motion. We should note that Carnot now uses the symbol $\sum$ to designate an addition. In other algebraic manipulations, an addition may be an integral or a summation.

Within this set of theorems, Carnot specifically treats the relationships between work and energy, with the denomination of work as a moment of activity, and the kinetic energy as living force. However, we selected the two following theorems in which he enunciates the principle of living force conservation and the principle of virtual velocities, respectively.

Theorem XXII

When a hard bodies system, free or constrained to any machine without spring, and animated by any driven forces, varying the motion by insensible degrees, the living force at the beginning of a given time is equal to the initial living force, plus the living force which would appear, if each one of the bodies' system would have only this velocity which it would acquire describing freely the curve which was described; supposing on the other hand that it was not animated at each point of this curve, only by the same driven force which it is actually submitted, and that its initial velocity has been zero.

The theorem above could be interpreted by Eq. (6.2). Carnot applies this equation to several particular problems, including work and living force of the gravity center of the machine.

Theorem XXIV

If some forces are applied to a system or to any machine maintaining mutually the equilibrium, and if this equilibrium was perturbed by the action of a new force infinitely small, the addition of the products of each one of these forces by its virtual velocity estimated in the sense of this force, i.e., by the velocity infinitely small from the point where it is applied, estimated in the sense of this force, will be zero.

6.2 Considerations About the Application of Driven Forces to Machines

Starting from page 227, paragraph 252, Carnot directly addresses the questions related to machines. It is in the final part of his *Fundamental Principles*, from which we maintain the same subtitle, that he studies the problems arising from machine operation deeply. The sequence followed by him starts with machines in equilibrium, its motion analysis, and the forms for increasing machine efficiency. Obviously these are problems that emerged from the needs of industry and engineering at that moment and therefore the polytechnician engineers of the first decades of the XIXth century since the publication of *Fundamental Principles* would develop new concepts and a new theory of machines. It is important to point out the role of the concept of work to these developments. As we will see, this concept occupies a central position in the majority of the investigations of this period.

Carnot starts by defining a machine as a body or a system of bodies which are in-between two or more powers (forces), with the condition of fulfilling a given objective. He emphasizes that, in general, these bodies are considered without mass because its small effect on the applied system of forces, independently if these forces are driven or inertial forces. He states that this abstraction simplifies the problem.

In this initial section, Carnot also discusses questions of modeling a machine through the use of a convenient representation with a great quantity of corpuscles separated by strings and bars, through which motion is transmitted from one element to its neighbors and so on.

Carnot is aware of the general character of his investigation and said that his intention was not to search for *particular properties of each machine, which he has remarked upon before, but to offer some considerations about the machines, and the common properties to every machine.*[10]

He then treats the problem of equilibrium and motion of a machine, noting the difference between the two types of effects caused by applied forces. For the equilibrium, only the intensity of forces is considered while for motion it is necessary to take into account the velocity of the point of application of each one of the forces and the path described by them. In other words, Carnot′s method for studying machine dynamics is based on work.

The difference between the two effects is created using the example of equilibrium and motion of a weight passing through a pulley. For the equilibrium condition, we need to maintain the weight in a given position. For the dynamic situation, the weight must be elevated to a given height, but it is possible to visualize the equilibrium as the particular case where the velocity is reduced to zero. This also characterizes that a limit was reached.

If we look at this problem in terms of virtual velocities, the two forces which act in equilibrium are in a reciprocal ratio of its virtual velocities, respectively, estimated in the sense of forces, and could be imagined as the lever's equilibrium. For the motion situation, we are dealing with actual velocities and not virtual as in the equilibrium condition.

The property which comes from the lever implies that a small force can maintain the equilibrium of a huge weight and which differs from the case of elevating a weight to a certain height, where the force goes down some meters if it is smaller than the weight.

For the case of motion, not only must the weight be considered, but the height to be elevated, making these two operations with the machine quite different. For the equilibrium, the machine could centuplicate the force effect while, for the machine in motion, we have an invariable quantity which is always the product of a force by the path described, estimated in the sense of force. In other words, Carnot discusses the exact question of energy in the field of mechanics, which is expressed by mechanical work. To eliminate any doubt, he uses an example of a horse which is two times stronger than another, meaning that it can elevate a new quantity of water, for instance, to twice the height as the second in the same time, or a double quantity of water to the same height, also in the same time. It is important to state that the definition of mechanical power appears here in this context in a very simple form. In the next chapter, this problem will be studied by Coriolis and other polytechnician engineers.

Another important question related to economic concerns is the capacity of a machine to elevate a given weight to a certain height. This is the method used to calculate the machine's capacity to provide work as well as the labour of workers

---

[10] Before Carnot, machines were studied case by case as a succession of particular cases. Carnot had the objective of a general theory of machines.

and its estimation of the capacity to pay salaries. This question will be discussed in the next chapter. What Carnot supposes is that the concept of work in physics is also useful for calculating a salary's value as referred to in the *Introduction*, although Carnot's concerns with economic questions also related to mechanics is a remarkable fact.[11]

In the context of the above discussion, Carnot enunciates the famous principle: *In any machine in motion, one loses always in time or velocity what one wins in force.* Carnot then analyzes the true meaning of this principle, discussing the effect produced by a machine in motion. If we call P the weight to be elevated to a given height H, this effect will be represented by PH. Assuming that the force used to produce that effect is F, V the velocity estimated in the sense of force, T the time during which the operation occurs, and supposing that, for the sake of simplicity, the motion is uniform, one has $FVT = PH$. As mentioned before, Carnot identifies the work through the moment of activity, which is equal to FVT.

And yet, continuing certain definitions, Carnot characterizes F as a driven force and P as a resistant force. What he calls inertial force is $\int m dV'/dt$, where m represents the mass of each one of the corpuscles of the machine. Because the motion is uniform, $dV' = 0$. Consequently, the moments of activity of these two forces F and P must be equal.

We have seen that $FVP = PH$. If a new force f with velocity u, acting during a time t, then $fut = PH$, or $fut = FVT$. The effect PH is independent of the kind of machine and cannot be overcome by the machine's capacity.

Carnot arrives at a discussion of a machine's utilization and states: *The advantage which machines present is not to produce a great effect with small means, but to provide the choosing among different means which we can consider equivalent, to the most convenient circumstances.* In other words, a machine is not a mere tool for the multiplication of forces, but mainly an apparatus with the availability of a certain amount of work that can be used in a great variety of forms. Carnot concludes this discussion by saying that it is always necessary that the moment of activity spent by the driven forces be equal to the effect of the motion absorbed at the same time by the resistant forces. These considerations seem to be sufficient to complete the illusion that machines having assemblies of levers *mysteriously* linked can transform a weak agent capable of producing huge effects, by transposing the reasoning adopted for the equilibrium situation to the motion condition. Indeed, for a machine in motion it is always limited and thus, the machine cannot overcome the spent moment of activity produced by the correspondent agent. The difference is for the equilibrium condition, that motion is obstructed; for the motion situation, the objective is to facilitate its appearance and, therefore, to

[11] A machine study, even one such as that conducted by Carnot with the purpose of knowing their motions and finding a physical approach to the problem related to forces, cannot be dissociated from the problem of replacing human labour. In other words, economic questions are automatically underlined.

maintain it. This condition necessitates one more consideration, which is to know the actual velocity of each point of the system.[12]

Carnot describes in detail, for the equilibrium and motion situations, the internal processes of machines related to the dissipation of motion by obstacles. For the dynamical case, the fixed points and obstacles, of any type, are forces of passive nature, absorbing motion of any intensity but never providing its birth, even for a small one, when the body is in equilibrium. Although a small power cannot annihilate a great power, only the resistance can be imposed by fixed points. In other words, a small power is only capable only of annulating a small part of a great one and the obstacles do the rest.

Carnot then studies the problem of the transformation of work into motion by considering all the parameters involved. From this viewpoint, it means establishing convenient variations of the terms of the quantity *FVT*, i.e., the moment of activity, later on denominated as work by Coriolis, as we will see in the next chapter. Thus, if time is the most important parameter and we should minimize it, the effect must be produced in a very short time. It is possible to generalize this reasoning for the case of a system of forces; for instance, if we have the forces F, F', F'' with the velocities V, V', V'', acting during the times T, T', T'', respectively, then one reads:

$$FVT = F'V'T' = F''V''T'' = PH.$$

If the motion of each one of the forces is variable, we will take the quantity: $\int (FVdt + F'V'dt' + F''V''dt'')$, or if we have the forces' directions with respect to velocities, one has:

$$\int \left[FVdt \cos\left(F \wedge V\right) + F'V'dt' \cos\left(F' \wedge V'\right) + F''V''dt'' \cos\left(F'' \wedge V''\right)\right]\cdot$$

This is the definition of work done by all forces.

The quantity *PH*, the effect to be produced by machine, is called the latent living force by Carnot. If we call M the mass of the weight P, and V the velocity correspondent to a height H, one reads:

$$PH = {}^1\!/\!_2 MV^2.$$

The relation above is always valid for any variation of the effect. When Carnot presented the above equation, he mentioned Leibniz as being its author, and said that only after Leibniz were the forces acting in bodies in motion calculated in a different form than the equilibrium situation.

With the following citation, Carnot anticipates the importance of the relationship of work to living forces, and, as we will see in the next chapter, this will be

[12] This discussion is at the center of the question of energy conservation and can be used in different forms, but is limited to a certain quantity establishing a limit to its capacity to undertake certain work.

the most important subject for the polytechnician engineers of the first decades of the nineteenth century: *For any denominations adopted, the consideration about what we call living forces will be always important to the theory of machines in motion, because are these forces which will be used to calculate men's work, animals' work, and of other agents that we can compare.*

Following Carnot's concerns, the problem of the efficiency of machines comes back. This is a question of central importance not only to mechanics of that time, but to the development of applied mechanics, and, later on, to industrial mechanics. Carnot remembers that all previous considerations about machines in motion were made without taking into account shocks and sudden variations of velocities. The variations considered are always by means of insensible changes; otherwise, we would have a great loss of living forces. He states: *In order to obtain from machines the best possible effect, it is important that it has been built such that the motion does not vary unless by insensible degrees. The exceptions are that with possibility to support different percussions, as the majority of mills. But even in this case, obviously we should avoid sudden variation unless that essential to the machine structure and operation.*

Carnot concluded the discussion above as to how to obtain the best possible effect in a hydraulic machine, moved by water flow, denying that the solution is to adapt a water wheel to the machine, because the shocks against water appear necessarily. Thus, two reasons can impede the maximum effect from being obtained: the fluid percussion itself and when this shock occurs, since there is always a residual velocity associated with the pure loss, and in some cases it could be employed to produce yet a new effect added to the first. To design an improved hydraulic machine, with the capacity to produce the maximum possible effect, we should look to fulfill the following conditions:

(1) For design considerations, to oblige that fluid could lose all motion by its action over the machine, or at least that the remaining motion be precisely that necessary to escape.
(2) That fluid completely loses the motion by insensible degrees without the occurrence of percussion from the fluid as well as from some machine components. This is independence of this type of machine.

The machine that better fulfills the conditions above will always produce the greater possible effect. Because it is so difficult to achieve the above objectives, particularly the water wheels, where shocks and percussions appear, it is at least possible to fulfill the second condition.

When Carnot discusses, in the following lines, what should be taken into account in order to produce the greater possible effect, he affirms that this problem depends upon particular circumstances and therefore the problem does not allow for a general solution to be applied for any situation.

The effect produced by a machine is a real or latent living force, always referring to the product PH of one weight P and a height H; we can call this effect q. On the other hand, to produce it, we need all driven forces to spend a moment of activity Q, which cannot be lesser than q; this means that there is no lost moment of activity to be consumed by a driven force, or that $Q = q$. But the moment Q of activity consumed by the force F in a time T, moving with velocity V, if we suppose a simplified form, F and V constants, and that the angle between F and V can be designated by *(F ^V)*, we have *FV*cos *(F^V)* as the quantity that must be maximized.

The equation above is, obviously, the moment of activity or, that is, the same as the work done by the driven force. It depends upon four quantities: F, V, T and *(F ^V)*; one form to maximize this product is to make it so that the direction of force coincides with that of velocity, i.e., that the force is in phase with the velocity, using other words. With respect to the quantities of force, velocity, and time during which the force acts, it is quite difficult to determine its intensities in an absolute form. If we could calculate it approximately, the problem of optimization should be tried.

Carnot comments on the case of work done by a man where fatigue is also involved. The knowledge of his physical constitution is fundamental. In general, this data could only be obtained by experience. We will return to this problem in the next Chapter in the study of Coulomb's contribution.

Within the work study undertaken by man considered as a machine which originated the fields of ergonomics and physiology, Carnot pointed out the importance of such previous studies as, for instance, Daniel Bernoulli's *Resultat de plusiers experiences destinées à determiner la quantité d'action que les homes peuvent fournir par leur travail journalier, suivant les differentes manières don't ils emploient leurs forces*; Bossut's *Mécanique* is referred to as containing important considerations about machines and a Euler memoir entitled *De Machine in Genere* is also referenced.

If we are looking at the maximization effect obtained by the machine, it is important to minimize the effect of passive forces, such as the friction force, string stiffness, air resistance, etc. Because it is impossible to eliminate all these resistances and passive forces, which impose the progressive decrease in machine velocity, this implies the impossibility of perpetual motion. If percussion does exist, the motion will decrease rapidly and the addition of living forces always decreases when percussion appears. In this context, it is important to quote Carnot concerning the impossibility of perpetual motion: *Obviously we cannot produce a perpetual motion, if it is true that all the driven forces that exist in nature are attractions and that this force have the general property, of being always the same to equal distances, between given bodies, i.e., a function that does not vary unless for the case where the distance of these bodies varies itself.*

Again, Carnot discusses the importance of the concept of work within the theory of machines, which is recognized by him as follows: *A general consideration about what was said, is that kind of quantity called moment of activity, plays a great role in the theory of machines in motion: because is in general this quantity that is necessary to save the maximum possible in order to obtain from one agent all the effect which can be done.*

If we are studying a machine at rest, where it is necessary to overcome bodies' inertia, but if we wish to give rise to a motion, the moment of activity which must be spent will be equal to the half-add of living forces which it should birth.

In conclusion, if we have any system of bodies animated by any driven forces with several external agents, such as men, animals, being employed to move this system in different ways, Carnot then enunciates the theorem of work and energy in the following form: *For any variation in the system, the moment of activity, which was spent in any time by the external powers, will be always equal to the half of the quantity of the addition of increasing of the living forces during this time, in the system of bodies where the powers are applied, minus the half of the quantity that was increased this same addition of living forces, if each one of the bodies was freely moved along the curve described by this body, supposing then that it has passed by each point of this curve, with the same driven force which actually acts.*

In Carnot's book *Fundamental Principles,* as analyzed above, the concept of work plays a fundamental role in the building of his new general theory of machines. In the next chapter, we will see how this concept was a starting point for other developments, mainly achieved by the polytechnician engineers. Once more, this theory is completely integrated into the conceptual framework of Rational Mechanics, because the tools used are based on its fundamental concepts, including the concept of geometric motion which is a modified version and an attempt to generalize the principle of virtual velocities. Therefore, the structure of *Fundamental Principles* contains a mechanical review emphasizing the problems involving shocks and sudden variations of velocities, which is precisely the transmission of motion inside the machine. The originality of Carnot's contribution is clearly different from Lagrange's mechanics, in the sense of continuous variations, or in insensible forms, as he characterized, but presenting a simple approach using geometry and trigonometry, where a cosine law is transformed into an equation of energy conservation.[13] Finally, the new theory of machines proposed by Carnot is also updated with the current practice of engineering of his time, which is shown in the last section.

Reference

Locke J (1973) Ensaio Acerca do Entendimento Humano, Coleção Os Pensadores, Abril Cultural, S. Paulo

[13] The fundamental differences between the mechanics of Lagrange and Carnot are that the first studies motion through continuous variations of position and the second through sudden variations, but the quantity known as work plays a fundamental role for both. Another difference between them is related to the methods of analysis. Lagrange uses variational calculus while Carnot applies geometry and trigonometry, which is a revalorization of the physical theories. Thus, geometry, as an ancient form of mathematics, found new possibilities with Carnot.

Chapter 7
The Metamorphosis of the Work Concept and its Incorporation into Economic Thought

First of all, the work is a process in which participate man and nature, and human being with his action propels, regulates and controls his material interchange with nature. He faces the nature as one of his forces. He moves natural forces of his body, arms and legs, head and hands, in order to appropriate of nature's resources, giving useful shape to human life.
In labour process, the activity of man operates a transformation, subordinated to a given objective, in the object in which operates, by means of the work's tools. The process finishes when product is made. The product is a use-value, one nature's material adapted to human needs through the shape change.

(Karl Marx (1968) The Capital, vol 1, Chap. 5, Editora Civilização Brasileira, Rio)

7.1 The Conservation of Living Forces Principle

As we have seen in the previous chapters, Lazare Carnot makes extensive use of the living forces conservation principle, which is equivalent to relating them directly to the concept of work, but also globally, as a balance over the whole system, as a general conservation principle. Let us see now a brief history of its application to machines.

According to Navier, the first study where we see the principle of living forces conservation applied to machines is *Hydrodynamics* by Daniel Bernoulli, published in 1738. It is also the first time in which the relation between hydrodynamics and that principle is established. Daniel Bernoulli shows that, if we abstract frictions, losses and we consider an incompressible fluid, an increase in the speed of the fluid occurs simultaneously with a decrease in pressure or a decrease in the fluid's potential energy, relating speed with height and pressure in a method similar that used in the elevation of a weight to determine the capacity of the machine. Bernoulli performed his experiments on liquids, so his equation is valid

A. R. E. Oliveira, *A History of the Work Concept*, History of Mechanism and Machine Science 24, DOI: 10.1007/978-94-007-7705-7_7,

for incompressible flow. Thus, in an arbitrary point along a streamline, the sum of living forces, weight times height and pressure divided by fluid density is constant.

Unfortunately, this important achievement made by Bernoulli, actually expressing an energy balance, was completely forgotten in the most important works on mechanics looking for an application, mainly those addressed to engineers. Among others we can cite *Physics* by Jean Théophile Désagulliers (1683–1744) and *Hydraulic Architecture* by Bernard Forest de Bélidor (1698–1761). Even the physicists and mathematicians that worked on more theoretical mechanics also did not realize the importance of Bernoulli's principle. Euler, for instance, did not use Bernoulli's ideas when he studied the reaction wheel, the centrifugal wheel and Archimedes' screw. Only with the memoir of Jean-Charles Claude Borda (1733–1799) entitled *Memoir on the hydraulic wheels*, published by the Sciences Academy in 1767, did the principle of living forces conservation begin to be known and to be applied to machines. Borda was the first to apply this principle to hydraulic wheels. He adopted Bernoulli's method, improving it on one point. While Bernoulli supposed that the living force lost by the shock effect was equal to $m(v^2 - v'^2)$, Borda stated that loss by the shock's laws would be $m(v - v')^2$.[1]

Some years later, in 1781, Coulomb published a memoir on the windmills and used the same principle to study this machine. In addition, he included the losses caused by shocks. Coulomb's text was entitled *Theoretical and experimental considerations on the effect of windmills*, also published by the Sciences Academy (Coulomb 2002).

To Navier, the contributions of Borda and Coulomb were fundamental steps and remarkable progress when these developments were compared to Bernoulli's *Hydrodynamics* in spite of their particular restricted applications of the living forces conservation principle. Navier also emphasized that, after these studies mentioned above, there was a need for the creation of a general theory involving that principle with the capacity to calculate the efficiency of machines. It was exactly in this context that he affirmed that this theory was created by Lazare Carnot in his *Fundamental Principles of Equilibrium and Motion*, which was analyzed in Chaps. 5 and 6, respectively. Navier attributes to Carnot the general demonstration of the theorem which calculates the loss of living force due to shocks between non-elastic bodies (hard bodies). This fact was not observed either by Borda or by Coulomb, except in very particular cases.

These considerations made by Navier and Carnot's recommendations for the purpose of increasing the efficiency of hydraulic machines were previously taken into account.

After these achievements, the person who made an important contribution to the development and application of the living forces conservation principle was Navier. In a note published by the *Proceedings of Chemical and Physics*, he

[1] If V is the approaching velocity of the fluid to a water wheel and V' is the fluid velocity, the second expression, proposed by Borda, calculates the kinetic energy which can be communicated to the water wheel.

communicated his intention to submit to the Academy some notes and additions prepared for a new edition of Bélidor's *Hydraulic Architecture*. It was a kind of update of this important book. In his remarks, Navier demonstrates the principle of living forces conservation to a single mass but he also generalizes this result to a system of n particles through the d'Alembert principle. This study was made by means of virtual velocities and Navier later derived Carnot's theorem from the conservation of living forces. According to him, this theorem fulfills the principle of living forces conservation for the case of a sudden change as follows: "The addition of living forces which arise in a system after a sudden change, is lesser than that which occurs previously, and the system has lost an amount of living force equals to that which would happen if the bodies were animated of velocities that are lost since this change".[2] As usual, Carnot uses the decomposition of velocities proposed by the d'Alembert principle before applying the living forces balance.

However, as we will see in this chapter, a great evolution in the living forces principle occured with the publication of Coriolis' book, *Du Calcul de l'Effet des Machines*, published in 1829 (Coriolis 1829). This work is considered one of the most important in mechanical engineering to appear in the nineteenth century. In addition, this book was a fundamental step in the history of work concept, because the term *work* is coined, becoming adopted by the polytechnician engineers as well as by the future technical literature.[3] Thus, the word *work* was progressively used and replaced the previous denominations such as mechanical power, quantity of action or mechanical effect, etc. As we have seen, Carnot called work the moment of activity.

If we compare the famous Coriolis work with Carnot's *Fundamental Principles*, it is easy to see that it represents great progress from the technical point of view, as well as through its completeness, style and language. The 25 years that separates these two fundamental books in the history of mechanical engineering did not contain important works, except the notes and remarks by Navier about Bélidor, as previously mentioned. This advancement in the mechanical field, which means the development of applied mechanics in the context of industrial progress, will be studied in detail in this chapter. This process, which we try to characterize along its general lines, mainly in the *Introduction*, had a

---

2 This theorem of Carnot's means a global energy balance over the whole system where all the components, including that of work, are converted into living forces.

3 In fact, the incorporation of the term work is more complex because not only did Coriolis adopt it, but other polytechnician engineers did the same, for instance, Poncelet. In Ref. (Poncelet 1870), he states: "This expression, mechanical work, which in some way is self-defining, I have used at the same time of quantity of action, in writing my course at School of Application in Metz (published edition, at the beginning of 1826 and presented in the same year to Academy of Sciences, which was readdressed to a Commission with the participation of Arago and Dupin)... but I did not adopt this expression mechanical work, in a definitive manner, unless exclusively in relation to any other, only in my lessons of 1827 to workers, after to be encouraged by Coriolis, who used in his repetitions at Polytechnic School, in a time before the publication of his Du Calcul de l'Effet des Machines, which appeared later on".

great influence on engineering education due to the need to prepare the technicians to give continuity to technical progress itself and thus also provide a fundamental requirement for the Industrial Revolution, which was, at that moment, spreading across the European continent. As we know, Coriolis was a lecturer at the Polytechnic School, Navier did the same at the School of Bridges and Highways, and Poncelet taught at the School of Artillery and Engineering in Metz.

In the next section, we will see that the work concept will be used not only in the mechanical field, but will acquire some other useful characteristics for machine science. These transformations mean that the concept initially applied to machines was enriched with an economic dimension, as referred to by Carnot in an attempt to measure work value performed by men and machines. This process has a great epistemological importance to mechanics and also to economy. This new content incorporated into the work concept can be considered a collective work performed by the polytechnician engineers.

7.2 Coulomb and the Work as a Form for Overcoming Passive Resistance

Alongside Lazare Carnot, Coulomb can be considered one of the most important founders of applied mechanics. Carnot, as is well known, created the first general theory of machines. Carnot's theory of mechanics emerged in the context of the development of Rational Mechanics under the influence of d'Alembert's principle. Coulomb, through the discovery of laws of friction, created a new science for studying the phenomenon of friction in machines. In addition, Coulomb provided solutions for many of the important problems of applied mechanics, such as his elegant solution to the problem of torsion in cylinders and his use of torsion balance in physical applications for physicists in successive years (Fig. 7.1).

Coulomb was the first to show how torsion suspension could provide physicists with a method of accurately measuring extremely small forces.[4] By using torsion balance he achieved remarkable results: the law of attraction and repulsion, electric point charges, magnetic poles, distribution of electricity on the surface of charged bodies, etc. Based on experimental investigations, Coulomb developed a theory of attraction and repulsion between bodies with the same and opposite electrical charges. He also demonstrated inverse square laws for such forces and went on to examine perfect conductors and dielectrics. He suggested that there was no perfect dielectric, proposing that every substance has a limit above which it will conduct electricity.

These works on electricity were also important for the mechanist view of the world because they confirmed for the field of electricity a similar law to Newton's

[4] Using a torsion balance, Coulomb obtained important results in electricity and magnetism: the law of attraction and repulsion of electrical charges, study of magnetic poles, distribution of electricity on surface of charged bodies, etc. See Oliveira (2004).

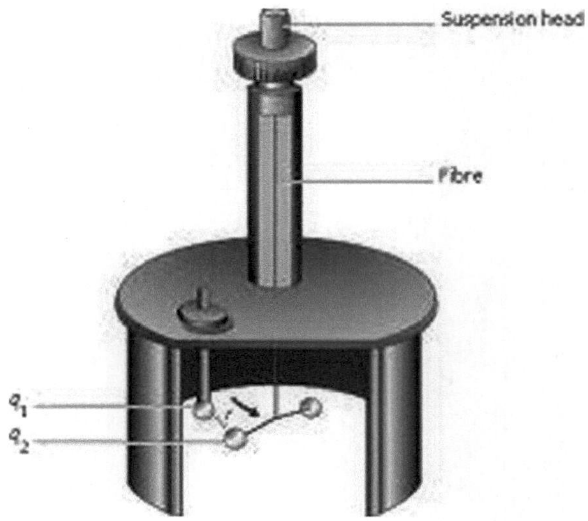

Fig. 7.1 Coulomb's torsion balance

theory of gravitation, both based on action at a distance between masses and an inverse square law of the same distances.

Charles Augustin Coulomb was born in Angoulême, France on June 14, 1736. His father was Henry Coulomb and his mother was Cathérine Bajet. Both his parents came from families well-known and important in their regions. After being brought up in Angoulême, the capital of Angoumois in southwestern France, Coulomb's family moved to Paris. In Paris, he entered the "Collège Mazarin", where he received a good classical education in language, literature and philosophy. He received the best available education in Mathematics, Astronomy, Chemistry and Botany.

After a period in Montpellier, Coulomb went to Paris in October 1758 to receive the tutoring necessary to take the examinations to enter the "École du Génie" in Mézières. He studied Camus's famous book, "Cours de Mathematique", for several months. In the same year, Coulomb took the examinations set by Camus himself. He passed and entered the "École du Génie" in February 1760, the same school from which Lazare Carnot graduated in 1773 (Fig. 7.2).

Coulomb graduated in November 1761, now a trained engineer with the rank of lieutenant in the "Corps du Génie". Over the next 20 years, he lived and worked in several places where he was involved in engineering, structural design, fortifications, soil mechanics and many other areas. After his first posting in Brest in 1764, he was sent to Martinique in the West Indies. However, Martinique was attacked by a number of foreign fleets over the following years and finally captured by England in 1762.

On his return to France, Coulomb was sent to Bouchain. He now began to write important works on applied mechanics and, in 1773, he presented his first work to the "Académie des Sciences" in Paris. This was entitled: *Sur une Application des règles, de maximis et minimis à quelque problème de statique, relatifs à l'architecture*. The most significant aspect of this work is Coulomb's use of the calculus of variation to solve the problem under consideration.

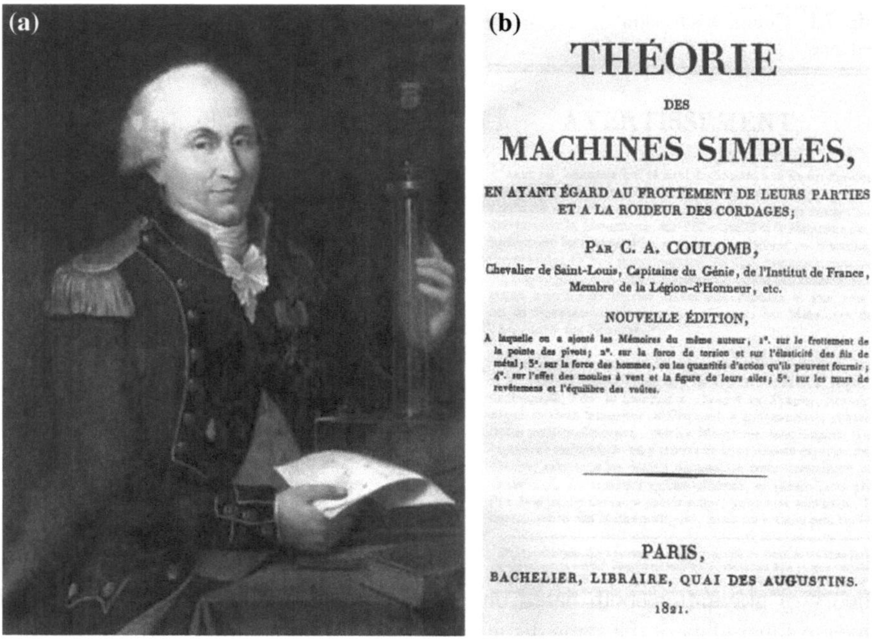

Fig. 7.2 **a** Charles-Augustin Coulomb (1736–1806), **b** Théorie des Machines Simples

In 1779, Coulomb was sent to Rochefort. During his time there he began his investigations into mechanics, using shipyards in Rochefort as laboratories for his experiments. His studies of friction forces led him to write the major work: "*Théorie des Machines Simples*".

This 1781 memoir completely changed Coulomb's life. He was elected to the mechanics section of the "Académie des Sciences", and moved to Paris where he now held a permanent post. He devoted the following years primarily to physical problems rather than engineering. He wrote treatises on electricity and magnetism, also submitted to the "Académie des Sciences" between 1785 and 1791.

He presented 25 *memoires* to the "Académie des Sciences" between 1781 and 1806. He also carried out several investigations with Bossut (1730–1814), Borda (1733–1799), de Prony and Laplace (1749–1827) during this period. He participated in the work of 310 Academy committees. Besides these engineering projects, he also undertook services for the French government in several fields, from education to hospital reform. His educational activities were largely carried out between 1802 and 1806, when he was inspector general of public instruction. He was also mainly responsible for setting up the *Lycées* across France. Coulomb died in Paris on August 23, 1806.

In the Introduction to his *Théorie des Machines Simples*, Coulomb referred to Amontons' work as follows: "Amontons, in his memoir to Sciences Academy in 1699, seems to be the first author that tried to measure friction and stiffness of strings regarding machines calculation. He believed, by means of his experiences that surfaces' extension has not influence in friction, and that friction depends only of the pressure between the surfaces in contact; he concluded that friction is proportional to pressures". Coulomb also refers the works of Muschembroek, Camus, Bossut and Desagulliers, as important contributions to the study of friction in machines.

Amontons performed the first experimental investigation of friction forces. This study changed the way friction was interpreted as a physical phenomenon. Before Guillaume Amontons (1663–1705), friction forces were treated either as a resistance to be overcome, resistance at the beginning of a relative motion between two bodies, or as a kind of adherence between surfaces in contact. In his famous study, Amontons showed that friction is independent of surface extension and stated: "If we think carefully about the nature of friction forces, we can conclude that it is an action in which a body under pressure from another can be moved through the common surface between them".

Amontons belonged to the founding group of scientists of the French Academy. In 1699, he published two memoirs, the first, "Moyen de substituer commodement l' action du feu à la force des homes et des chevaux pour mouvoir les machines", deals with a rotating machine that uses hot air as a kind of "fire windmill". The second is called "De la resistance causeé dans les machines tant par les frottements des parties qui les composent, que par la roideur des cordes qu' on y emploie, et la manière d' en calculer l'un et l'autre". It had the same objective that Coulomb proposed to solve later. In other words, the latter investigations attempted to understand the nature of friction forces and separate the effect of bending stiffness and friction. To do this, Amontons used various materials, such as copper, iron, lead and wood, in his experiments. Unlike Coulomb, however, Amontons used lubricating substances between the samples with relative motion, and established rigorous control over two parameters, the pressure between bodies and the force to maintain motion. Obviously, he made variations in the sample material. The experimental set up used clearly identified that the phenomenon was strongly dependent on the weights involved, as well as the nature of common surfaces in relative motion. Amontons did not publish numerical results of this investigation but summarized his achievements as follows:

(1) The resistance caused by friction forces varies in the same proportion as pressures between the parts in relative motion of more or less the same size.
(2) The resistance caused by friction forces is nearly the same in iron, copper, lead or wood for any variation introduced once the same lubricant material is used.
(3) The resistance of friction forces is approximately one-third of pressure.
(4) The resistance of friction forces depends not only on the weights and pressures involved but also on time and velocities.

The above postulations can be seen as Amontons' laws of friction. Obviously, this study influenced Coulomb's investigations. As shown below, Coulomb introduced a great deal of rigour into his study of friction forces.

With respect to the force necessary to bend a rope around a cylinder, Amontons showed that it was inversely proportional to the radius of the cylinder and directly proportional to the tension and the diameter of the rope.

Let us now analyze Coulomb's theory of friction. The complete title of Coulomb's major work is: "*Théorie des Machines Simples, en ayant égard au Frottement de leurs Parties et a la Roideur des Cordages*". It has 254 pages and is divided into two parts. The first part, with which we are particularly concerned, has 99 pages and is subtitled: "*Du Frottement des Surfaces Planes qui Glissent l'une sur l'autre*". It is divided into two chapters: Chapter I is entitled "*Du premier effort pour vaincre le frottement, ou pour faire glisser une surface après un temps de repos donné*", and Chapter II "*Du frottement des surfaces en mouvement*". Obviously, Chapter I deals with static friction and Chapter II with the dynamic case. With respect to the first case, Coulomb discusses the friction dependence of four types of causes:

(1) The nature of surfaces in contact and their surface finishing
(2) Surfaces extensions
(3) The pressure between the surfaces in contact
(4) The elapsed contact time between the surfaces.

The experimental apparatus used by Coulomb is shown in Fig. 7.3 and is suitable for sliding pairs of surfaces by means of a weight and a cable passing through a pulley. The table, which supports the moving parts, is stiff enough to maintain the whole system isolated from the ground. Coulomb made the following remarks after three tests:

> We have seen in previous tests that friction resistance was smaller after one second in contact than after one or two minutes; but after one or two minutes, friction reached the maximum possible value. After this result we tried to determine a rate between pressure and friction as soon as friction limit, or the maximum growth of friction, was reached; we obtained the following values for this ratio:
>
> First test - - - - - - - - - 74/30 - - - - - - - - - - 2.46
> Second test - - - - - - - 877/406 - - - - - - - - 2.16
> Third test - - - - - - - - 2474/1116 - - - - - - - 2.21
>
> As the tests indicated, an approximate contact rate between pressure and friction force is obtained. This in spite of large differences in pressure, while I have tried to reduce the area between the surfaces to verify if this ratio is still the same.[5]

The three tests were made with samples of wood without any kind of lubricant between the pieces in contact and changing the dimensions of the samples, the contact pressure and the elapsed time at rest. The numerical values shown in the experiments refer to the weights of samples in pounds.

Looking at the dynamic case, the same experimental apparatus was used, the contact areas were reduced and, after several tests, Coulomb concluded that friction is independent of relative velocities between the contact surfaces, as well as the magnitude of their contact areas.[6]

[5] If Coulomb had measured the inverse relation, he would have found what is now known as the friction coefficient being a positive quantity lesser than one. This finding was obtained by Euler. See Merlet (2004)

[6] It is the measurement of the kinematic friction coefficient which is less than the static friction coefficient and stays practically constant with the velocity, unless this velocity becomes very high and modifies the mechanical properties of the materials in contact and in relative motion.

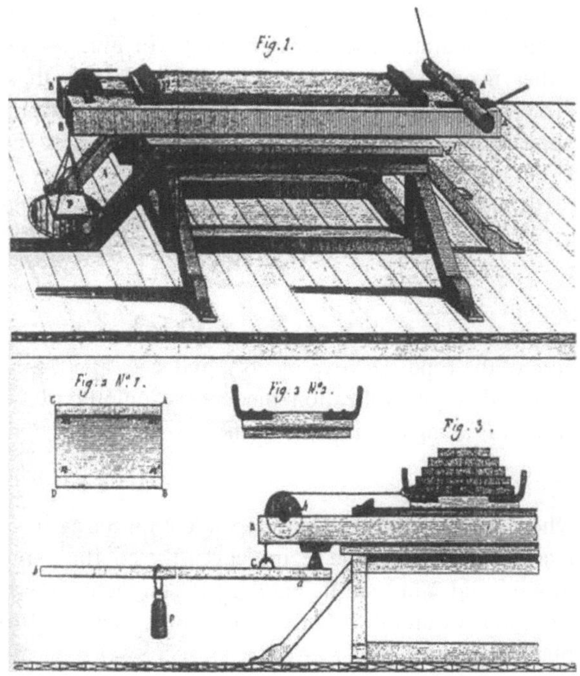

Fig. 7.3 Coulomb's experimental apparatus

In Chap. 3, Coulomb summarizes the results of these investigations as follows:

(1) With two pieces of dry wood in contact, after some time together without relative motion, the resistance increases significantly at the beginning before the relative motion arises; after some minutes the friction reaches its maximum, which is the limit for resistance.
(2) Since relative motion initiates with two pieces of dry wood, friction is still proportional to pressure; however, the intensity of friction is smaller than in the static case.
(3) If we look at metal surfaces sliding in relative motion without any lubrication material, friction is proportional to pressure; but the intensity of friction is the same, irrespective of how the motion is obtained after some time at rest.
(4) Heterogeneous materials in relative motion such as wood and metal without any lubrication substance lead to different results; friction intensity in relation to time at rest increases slowly and reaches a limit after 4 or 5 days.

This summarizes the well-known laws of friction,[7] laws as we know them nowadays. In the second part of his book, Coulomb studies the forces needed to bend ropes and friction in rotating shafts. The experimental structure consists of two suspended weights on each side of a pulley. In this type of device, forces to compensate friction effects are added to forces to bend rope elements. Obviously, it is

[7] These conclusions of Coulomb's were achieved after a very rigorous investigation into friction phenomenon and represent his laws of dry friction.

necessary to separate both effects. Coulomb uses the same method and the experimental apparatus used by Amontons. In Fig. 7.4, Amonton's machine is shown. Other investigations were performed, such as that carried out by Desagulliers (1683–1743) and published in his "Cours de Physique Experimentale", two vols, appearing in 1751. Basically, it was the same experiment as Amontons and Coulomb had conducted.

Coulomb generalized the effect of stiffness of ropes by means of the formula:

$$\frac{A + BT}{R}$$

where $A = h\,r^q$, $B = h\,r^s$, R is the radius of the pulley, r the radius and T the tension of the rope. The exponents q and s are approximately equal.

Another generalization made by Coulomb refers to the force T to move a weight P along a horizontal plane:

$$T = A + P/u$$

where A is a small constant dependent on the "coherence" of the surfaces and u is a coefficient that is the reciprocal of the coefficient of friction, which is now commonly used. This coefficient depends on the nature of surfaces. For a more general situation, Coulomb calculated the force necessary to hold a body on an inclined plane (Galileo 1987).

In spite of many developments in several countries, applied mechanics and further developments of industrial mechanics received an important heritage from the French polytechnic engineers (polytechnicians), associated with the knowledge of machine constructors since the end of Middle Ages in Europe. The successive creation of engineering schools in France, such as the School of Bridges and Highways (1747), the Royal School of Mézières (1748), the School of Mines (1781) and finally the Polytechnic School (1794), provided a homogeneous and high level of scientific education. In this context, the contributions of Coulomb and Lazare Carnot appeared as fundamental for the development of a new science of machines. As mentioned above, in addition to his proposed laws of friction, Coulomb solved several other important problems. Carnot created the first general theory of machines, thereby providing a tool for analyzing any kind of machine.

Coulomb is also responsible for the introduction of economic studies in relation to machines and can be considered a precursor of industrial mechanics.[8] This task was done by mechanically measuring the human body as a kind of machine in order to measure a machine economically. The memoir *Sur la Force des Hommes* is an attempt to understand human work mechanically and is the first publication on physiology and ergonomics.[9] The question introduced by Coulomb appears

---

[8] We cannot forget the question of machine efficiency, which was a general concern of that time, and the losses by friction were an essential part of it. With the study of friction laws, Coulomb made a significant advance towards the best working regime of machines, because friction appears in all of them.

[9] By using a mechanical analogy in order to try to measure the physical work of a man during a day of work, Coulomb returned to the case of a real machine and the question of its efficiency.

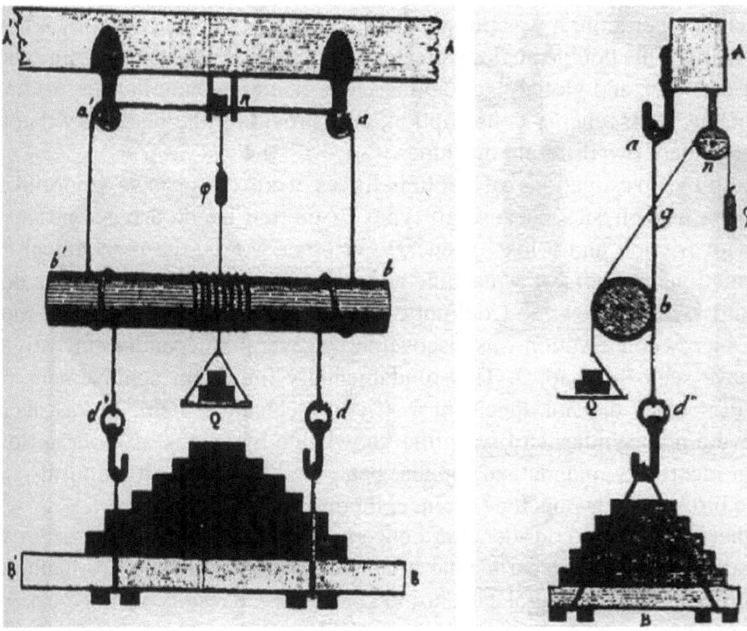

Fig. 7.4 Amontons's machine

from a pragmatic point of view. He proposed to measure the quantity of action (work) that a man can expend in a day of work by different ways of applying the forces. To do this, Coulomb studied an old problem of human mechanical capacity, also studied by Desagulliers and Daniel Bernoulli (1700–1782).

The heritage of Coulomb and Carnot was subsequently incorporated, developed and transformed into applied and industrial mechanics, mainly by the French Polytechnicians, Navier (1785–1836), Coriolis (1792–1843), Poncelet (1788–1867), Augustin Cournot (1801–1877), etc.

Augustin Cournot is known as a founder of probabilistic thought and also a precursor of the mathematization of economic theory. Before attempting to produce a broad synthesis of contemporary knowledge, he worked on mathematics and mechanics, being a disciple of Poisson. In 1834, he translated both Herschel's *Treatise on Astronomy* and Lardner's *Elements of mechanics* into French. After this period as a mathematician and expert in mechanics, he moved into the application of probabilistic methods to natural sciences. Finally he dedicated himself to the philosophical and economic analysis of machine science.

Coulomb's work is closely related to the beginning of the Polytechnic School education, especially the improvement in and new kind of teaching of the machine sciences. This change was conceived by Gaspar Monge (1746–1818) and his disciple Pierre Hachette (1769–1834).

In 1821, Coulomb's mechanical studies were published, including the "*Memoire sur la force des hommes*". By introducing the question of human work as a machine operation, it was possible to study the work carried out by a machine economically. This double analogy created the conditions for constructing the concepts of net work and global work applied to a given machine in order to measure its efficiency, costs, energy consumption, etc., providing the necessary theoretical tools to compare two different machines.

Coulomb's investigations of friction forces were of decisive importance for engineering and physics because all types of motion inside a machine are associated with friction and a loss of energy. In other words, the mechanical energy transformations by parts of a machine and its communication of motions degrade that quantity yet unknown to Coulomb—energy. It was only in 1847 that the principle of energy conservation was discovered by several independent investigations, as we have seen in Chap. 3. This fundamentally important achievement for the history of science did not appear in a strictly mechanical field. It was necessary to achieve a new synthesis of scientific knowledge by means of modern developments in electricity, magnetism, optics, heat, etc. This new situation also represents the birth of a new machine science: thermodynamics.

Another remarkable consideration concerns Coulomb's method. It is, in fact, a combination of in-depth experimental research with the application of mathematics, mainly differential and integral calculus, to generalize the results obtained experimentally, by means of simple algebraic formulas. This was done, for example, for friction in bending of ropes and in rolling. From an examination of many physical parameters, he developed a series of two-term equations. The first term was a constant and the second term varied with time, normal force, velocity, or other parameters.

The famous scientist Jean-Baptiste Biot (1774–1862) wrote: "It is to Borda and to Coulomb that one owes the renaissance of true physics in France, not a verbose and hypothetical physics, but that ingenious and exact physics which observes and compares all with rigor".

7.3 Navier and Work as a "Mechanical Currency"

Louis-Marie-Henri Navier, official of the honor legion, member of the French Royal Institute and inspector of the division of the *Corps des Ponts et Chaussées*, was born in Dijon, on February 15, 1785. At 14 years of age, he became orphaned from his father and found in his uncle Emiland Gauthey a kind of second father. Gauthey was a civil engineer who worked at the *Corps des Ponts et Chaussées* in Paris. He was considered the leading civil engineer in France and he certainly gave Navier an interest in engineering. Despite encouraging Navier to enter the *École Polytechnique*, Gauthery seems not to have been that successful in teaching Navier (Fig. 7.5).

In 1802, Navier was ready to be submitted to examinations for *École Polytechnique*, being one of the first to enter with merit. After 2 years of studies in that school, showing remarkable performance, he also decided to enter the *École des Ponts et Chaussées*. In 1808, he became an engineer. Following the courses in the two schools of engineering, Navier acquired a great theoretical and practical ability. He directed the

Fig. 7.5 Louis-Henri Navier (1785–1836)

construction of bridges at Choisy, Asnières and Argenteuil in the Department of the Seine, and built a footbridge in the Île de la Cité, a famous place in Paris.

In 1819, he was designated as assistant professor of applied mechanics at the École Royale des Ponts et Chaussées. In 1824, Navier was admitted into the *French Academy of Science*, and in 1830, he achieved the position of titular professor of the *École Nacionale des Ponts et Chaussées*. In the following year, he succeeded Augustin Louis Cauchy as professor of calculus and mechanics at *École Polytechnique*.

Navier made a fundamental contribution to the theory of elasticity which was mathematically formulated in 1821, making this science available to engineers in different fields with sufficient accuracy. In 1819, he determined the neutral line of mathematical stress and thus corrected some of Galileo's incorrect results. In 1826, he established the elastic modulus as a property of materials independent of the moments of inertia. Due to these contributions to engineering sciences, Navier is considered to be the founder of modern strength of material and structural analysis. That same year, the first edition of his book on strength of materials appears in which his main achievements in that field were incorporated.

Perhaps the most important contribution to science made by Navier s are the Navier–Stokes equations, because of the participation of George Gabriel Stokes (1819–1903) in their establishment. These equations describe the dynamics of a fluid applying Newton's second law. This is done with the assumption that the fluid stress is the sum of the diffusion viscous term, proportional to the gradient of velocity and the pressure term.

The importance of these equations stems from their description of the dynamics of many natural processes and may be used to model the weather, ocean currents, water flow in a pipe, air flow around a wing, and many other applications. In addition, these equations can help to design aircrafts and cars, to study blood flow, to analyze pollution, etc. Associated with Maxwell's equations, they can also be used to model and study such new physical fields as magnetohydrodynamics, for instance. Lastly, from the mathematical point of view, the Navier–Stokes equations are a source of research leaving many open problems for future investigation.

Navier's contribution to the elaboration of the economic-physical concept of work comes from a short text entitled: "On the principles of the calculation and the establishment of machines and on the motors". This study also comes from the notes and additions written for the reediting in 1819 of Belidor's *Hydraulic Architecture.* Belidor's book, initially published between 1737 and 1739, became the most important reference for hydraulic engineers of that time. Obviously, regarding the date of publication, this handbook already presented old notions and concepts. In 1781, Coulomb announced a review of the handbook but it was Navier that accomplished Coulomb's project.

Navier practically wrote an entire second book as a kind of appendix to the original Belidor book, constituting a complete review. The approach that he adopted was to add notes and remarks to the original text instead of discussing the concepts. Through this method, he essentially wrote a group of treatises of sorts on the important questions that were used later as independent works. The concepts in which we are particularly interested appear in one of these additions.

Firstly, Navier uses the concept of work as a measure of machine production, which he denominates as *mechanical currency.* Then, he studies the question of machine efficiency and its maximization in the same sense proposed by Coulomb. In fact, the problem studied by Navier is the same faced by Carnot: the elaboration of a methodology for evaluating the efficiency of machines and motors. The basic difference between Navier and the previous investigations is that he analyzes the problem of machines in a more general context. incorporating economic questions.[10]

The fundamental question faced by Navier in that moment was to find a measurement tool to be used in production involving a mechanical system. To Navier and the polytechnician engineers, a machine was a system used to transmit forces in opposition to a motor considered an agent that produces forces. In order to achieve his objective more easily, Navier restricts his field of investigation to machines which are "submitted to a permanent action of a motor, performing a continuous work, with its different parts animated in a uniform motion or varying motions, but periodic in that the velocity has a constant medium value". This manner of classifying a machine by means of the complexity of its motion had already been used by Carnot in his Essays, as we have previously seen.[11]

[10] In fact, Navier's main contribution was reviving the question of work as an economic concern, where his designation of work as mechanical currency is perfectly suitable. From the physical point of view, there is no relevant advance to the physical concept. Its use is the same as that presented by the physics of that time.

[11] This simplification makes the driven work approximately equal to resistant work and the space displaced by force equal to the product of the velocity by the time.

As we will see, the concept of work does not appear as a clearly defined physical concept, but as a notion stemming from to common sense as far as its economic meaning stands. In this sense, work means the product, or what is obtained by a given production. According to Navier: "The comparison among several machines by the seller or by the capitalist is done naturally after that a quantity of work which they undertaken and the price of this work is established. To estimate the respective values of two flour mills, for instance, we will examine what quantity of flavor each one can mill in one year. In order to compare a flour mill with one of sawing, we estimate the value of the first after the quantity of flavor produced annually and the price of the established milling, and the value of the second after that the quantity of wood which it will produce in the same time and the price of sawing be defined".

For certain more complicated cases, Navier realized that an ordinary economic measurement was not sufficient to establish a satisfactory comparison. He discussed one of these cases in the following form: "Let us suppose that one person has a flour mill, and he wish, by means of some modifications in its mechanism to transform it in a sawing mill. The person cannot judge the advantage or disadvantage of this operation unless he knows how to evaluate later that the quantity of flavor produced by one mill will represent in terms of the quantity of wood that he will have. Or this evaluation is an absolutely impossible thing, unless he finds one common measure to the two different works".

The problem clearly proposed by Navier is how to find a tool or a measurement which permits an a priori calculation of this equivalence between the two types of work, for the case of the above example, the equivalent in wood of the flavor produced and vice versa. What is searched for measurement is the work capacity of a machine in the context of production, independent of the work nature in order to have an available common measurement not submitted to the market price. Using Navier's words: "To establish a sort of mechanical currency, if we could express like that, with which we can estimate the quantities of employed work to accomplish any kind of fabrication".

At this moment in the development of his analysis, he uses mechanical theory. Navier considers that mechanical science can provide a common measurement which is missing in the economic analysis. According to a current vision of that epoch, but also common to certain authors, he believed that all production consists in overcoming a mechanical resistance, i.e., by displacing a force or by deforming a body. As a complementary idea to this vision of evaluating a given work, we can adopt the general model of elevating a heavy weight, as already done by Carnot. Thus, Navier states: "There are always in a machine action, an effort of a pressure exerted against a point, while a displacement by a point in space is observed. This observation leads naturally to recognize that a kind of work the best suitable to be used to evaluate any kind of them is the vertical elevation of heavy bodies. Although, independently of his susceptibility to be expressed by a numerical precise form, without arbitrary things, we can ever, for any kind of undertaken work by a given machine, not only by thought or by an abstraction of spirit, but in reality, to replace this work by a weight elevation... The elevations of weights will represent then a machines's work and a machine will be considered as undertaking more work if it could elevate more weights to a bigger height".[12]

Navier has adopted the mechanical work to his mechanical currency, the definition of which was clearly done in the eighteenth century, obviously without use of this designation. He adopted the same expression for work as that used by Coulomb, that is, quantity of action.

However, if the problem studied by Navier seems to be solved from the physical point of view, it continues to be unsolved from the economic point of view. It is worth observing that, to find a solution, Navier simplifies the problem considering that it is independent of time. From the physical point of view, it is possible to abstract the time in the work definition, but from the economic point of view, that is impossible, as we will see. In general, the problem can be outlined as follows: how to compare two machines which can elevate the same weight to the same height, but with a time of operation that is doubled from one to the other?[13]

In the next section, we will see that Coriolis analyzes this problem in depth. Navier, after tackling the first part of the problem, which was to establish a measurement for the equivalence between different works, proposed the concept of mechanical currency, as we know. He then tries to study the problem of machine efficiency. He states: "The action exerted by motors on the machines to put them in motion and working, must be estimated in mechanics in the same way and with the same kind of unity of the work performed by machines. However, the motor acts on the machines how it acts on one resistance: there is ever in the point of application of the motor a resistance, a pressure exerted and displaced space".

Navier now discusses the concept of work in an economic context as follows:

> It will be worth in order to show because the quantity of action spent by work is considered as providing the true measure, to observe here that is always proportional to that quantity of action that are established the prices cash paid by different kinds of work. Indeed, when we pay a work, in fact it is the time of the worker what is paid; only this time is estimated more or less expensive, and work demands to worker more or less vigor, intelligence, or acquired knowledge. Or, if we conceive one worker employing his forces in a constant and regulated manner, he will exert constantly the same effort acting with a constant velocity, and consequently he will produce quantities of action equal in equal times. Then, the price of one work is proportional to the time necessary, being also the quantity of action that it represents.

In order to highlight the notion of useful effect, it was necessary to overcome the ambiguities yet present in the definitions regarding the relation between driven work and resistant work. Then, Navier uses the assumption previously made of a machine with uniform motion. In this case, if we neglect the starting and stopping

[12] Again, as with his antecessors, the problem of the capacity to undertake work of a given machine emerges. This returns things to the energy question and the discussion conducted by Carnot in the final part of his *Principes*.

[13] From the economic point of view, that is, for the purpose of comparing the work of machines, the parameter time is essential, because the costs in general, and primarily the work productivity, are directly related to the economy of time. And yet, as we have seen in the introduction to this book, a saving of time implies cost devaluation of human labour.

times, we could admit that driven work is equal to resistant work, because with constant velocity there is no loss of living force. In other words, the production is equal to the spent or expenditure. Thus, we can ask how the net effect appears. Navier tries to explain the notion of resistant work in a more detailed way, distinguishing useful work and lost work. By doing this, it appears that the notion of loss and the ambiguities disappear. Navier states: “It is necessary now to examine in a particular way the idea that we should aggregate the word resistance employed below. It appears naturally and is an obstacle to the machine motion, producing work that it must perform. However, it is important to observe that there is no machine that we can conceive, that does not overcome several obstacles to motion, independently of the type that we speak… It is necessary, then to any machine to conceive the pressure exerted by motor and shared in two parts, where one produces the equilibrium, the resistance itself performing work to be done, and the other the resistances from the machine; the quantity of action spent by the motor in its application point and shared also in two parts, where one is consumed in pure loss by the last resistances and the other product which we denominate usually useful effect of the machine”.

With the work’s components well defined, it is possible to define efficiency and thus to use it to compare machines. Using Navier’s words: “We can see then that the quantity of work performed by a machine, or its useful effect, is one part of the quantity of action provided by the motor, which cannot be overcome or even equated. One machine approaches each times more of the perfection depending of its useful effect approaches of the quantity of action consumed by it and is mainly toward this point of perfection that its working must be directed. The mean to obtain this in general is to design the mechanism as simple as possible and to avoid all shock among hard bodies and all sudden change of velocities”.[14]

It is easy to see that the last phrase of the citation above is the same recommendation made by Lazare Carnot in order to minimize machine losses.

Navier again discusses the problem of machine efficiency in the economic production context as follows: “The true objective in our machine’s study is the return in money which is looked for, that is proportional to its useful effect, to be the greatest possible in relation to the motor’s expenditure. Or the useful effect, always lesser than the quantity of action provided by the motor, increases or decreases with it. It is necessary then to proceed such that the quantity of action provided by the motor has a minimum possible cost, or to use the greatest possible quantity of action provided by the motor”.

The Navier text from which we took certain quotations does not present any new element from the physical point of view. The text does not have the objective of defining the concept of work within the framework of rational mechanics, but only of using this concept as a heritage of the eighteenth century. Its main objective is to show that the quantity of action or the work is the fundamental concept

[14] The problem of machine efficiency or the calculation of their mechanical efficiency is developed and becomes clear for Coulomb’s proposal studying human body as a machine, taking into account the associated fatigue. Navier’s proposal for improving machine efficiency is of a physical nature and quite similar to what was suggested by Carnot, as we have seen.

for measuring the product as well as the machines' expenditures. In this sense, Navier advances in relation to Carnot, being more profound in the utilization of the work concept, eliminating ambiguities, and mainly by introducing the mechanics of machines into the economics production framework. By doing this, he is one of those responsible for the process of incorporation of the physical concept of work into economic thought.

7.4 Coriolis and Work as a Measurement of a Machines' Operation

Coriolis was born on 21 May 1792 in Paris to a small aristocratic family that was impoverished by the French Revolution. The young Coriolis showed remarkable mathematical talents early. At 16, he was admitted to the Polytechnic School where he later became a teacher. During the 1830 Revolution, Agustin-Louis Cauchy (1789–1857) left France for a period and Coriolis was invited to succeed the famous mathematician.[15] In 1832, he became Navier's adjoint professor at the "*École des Ponts et Chaussées*". In 1838, he became director of the Polytechnic School, succeeding Pierre Louis Dulong (1785-1838) and accumulating activities of teaching and research. Coriolis died in 1843 at the age of 51 in Paris (Fig. 7.6).

At that time, the teaching of mechanics in engineering schools in France was dominated by statics such that problems of mechanics were related to constructional aspects not applied to machines. On the other hand, Lagrange's mechanical theory was criticized mainly because the difficulty in applying it to practical problems. At the same time, a movement was developed in France to improve the education of workers, craftsmen and engineers in rational mechanics. Coriolis was among the first to promote this educational reform, and in 1829, he published *Du calcul de l'effet des machines*, which later became a textbook with the objective of being used by industry engineers.

In this book, he corrected the expression for kinetic energy, introducing the constant ½ in the old expression mv^2 and adopting the term "work" to represent the product of a force by its displacement along the force trajectory, replacing the previous nomenclature. In fact, a rigorous analysis shows that the new term was adopted by the new generation of polytechnic engineers. In addition, Coriolis suggested a new unit for work called dynamode, but it was not adopted. The term dynamode is a composition of two words, "dynamis" meaning power and "hodos" to designate trajectory; 1 dynamode is equal to 1000 kg.m.

Coriolis lived in a period of great development in mechanics and science in general. At the end of the eighteenth century, in 1781, Coulomb presented to the "Academie des Sciences" his "Théorie des Machines Simples", creating a new

[15] Indeed, Cauchy is a scientist and remarkable personality, having been a prominent figure in France between 1815 and 1830. All mathematical and French mathematical-physical studies of the nineteenth century were greatly influenced by his studies, methods and even his style.

Fig. 7.6 **a** Gaspard-Gustave de Coriolis (1792–1843), **b** Du Calcul de l'Effet des Machines

theory of friction. Lagrange (1736–1813), in 1788, published his most important book, "Mécanique Analytique". Lazare Carnot, in 1803, as we have seen before, published his "Principe Fondamentaux de l'Équilibre et du Mouvement". Pierre Simon Laplace (1749–1827) and a group of scientists tried to generalize the conceptual basis of Newtonian mechanics to other particular sciences in what became known as the "Laplacian project". Heat phenomena was now beginning to be explained with the publication of "Théorie Mathematique de la Chaleur", by Joseph Fourier (1768–1830) in 1822, and a new science of machines appeared with Sadi Carnot (1796–1832), son of Lazare Carnot. His "Réflexions sur la Puissance Motrice du Feu", published in 1824, inaugurated thermodynamics.

The context illustrated above should be enriched by the economic ideas which appeared in the same period. The moment that the concept of work arose was the same as that in which the *labour-value* appeared by Adam Smith (1723–1790) and David Ricardo (1772–1823), the founders of political economics. Obviously, the concept of work was behind the ideas of some polytechnician engineers, but its genesis cannot be completely understood within the classical framework of machine sciences. The concept only emerged when the machine was thought of as an economic tool with the possibility of replacing man in productive function. This representation already appeared in Amontons (1663–1705) at the end of the seventeenth century when he proposed a common measurement for human and mechanical action. Thus, to measure a machine economically, it is necessary to measure human effort mechanically. This problem was discussed at the end of the eighteenth century by Coulomb, giving rise to the new sciences of physiology and ergonomics, as we have previously seen.

Besides the progress in science, the Industrial Revolution was spreading rapidly to Europe. Some ideas used in physics were then in current utilization in economics and vice versa. The concept of work belongs to this category, appearing as a bridge-concept to economics in a kind of incorporation and acquiring an economic dimension.

To engineers and economists, the most important question was how to measure the economic production and costs of machines and men working and operating together. Other fundamental and associated concerns were how to optimize these factors in an overall operation.

It seems that polytechnic engineers of this period were influenced by the economic ideas of Jean- Baptiste Say (1767–1832). His economic theory considered that production is based on the concept of utility, meaning the capacity of products to satisfy the consumer's needs. To him, the economic values were established by the market according to their utility. The concept of utility was central to Say's economic ideas and allowed for the development of a technical analysis about machines. On the other hand, Say's economic theory was strongly supported by technology and incorporated certain ideas about nature. To Say, work generated by machines is similar to labour because this can be replaced by machines. Thus, these ideas can perfectly express the polytechnic engineers' thinking. This situation is probably real. In 1819, Say began to teach a course on industrial economics at CNAM (Conservatoire National des Arts et Métiers), a famous French institution founded in the same year as the "*École Polytechnique*" (1794).

Before analyzing Coriolis' contribution to the incorporation of the concept of work into economic thought, it is necessary to mention his other important achievements in mechanics. Coriolis discovered the effect coined Coriolis' force, in 1831, as the result of an engineering concern.[16] This was made through the application of the living forces conservation principle to the relative motion of machines, especially to hydraulic wheels and turbines. He describes this process as follows: "The principle of living forces extended to relative motions provides easily an exact theory to hydraulic wheels similar to Borda or Burdin's turbines. To wheels with Poncelet's curved vanes, he shows that every time that water left the vane from the same distance of the rotation shaft where it came in, if we neglect frictions, the vane cannot acquires or loses unless the relative velocity due to gravitational effect, related to wheel considered at rest; although with the

[16] For a more detailed description of how Coriolis found the effect that bears his name, see the original text (Coriolis 1832), as well as the work of Koetsier (2003), in which he shows that Coriolis was studying a vertical water wheel with curved blades when that effect appeared. Coriolis' effect appeared through a balance of kinetic energy (living forces), considering the water relative motion entering vertically into the water wheel through the big diameter and leaving through the small one.

ordinary shape of vanes, the water relative velocity is grater by lefting the vane than in the entrance".

Coriolis published more two memoirs applying the principle of living forces to machines. One, in 1832, was entitled *Sur le principe des forces vives dans les mouvements relatives des Machines*, and was read in the Academy of Sciences in June 6, 1831. The second, in 1835, was entitled *Sur les equations du movement relative des systems de corps*. Both studies were addressed to hydraulic machines, analyzing the problem of fluid motion through channels in rotation. In other words, Coriolis generalized certain particular problems previously studied regarding relative motions to systems in rotation.

In 1835, Coriolis wrote *Mathematical theory of billiard game* (Coriolis 1990), with the main purpose of calculating the influence of shock on the mode of rotation and translation of a billiard ball, besides trying to predict the complex motion that would appear by means of the ground friction and the reactions of the balls.

Coriolis's textbook is the result of writings accumulated since 1819. His main objective was to develop the Lazare Carnot project which created a general theory of machines based on the d'Alembert principle, but, in general, as a product of the tradition of Newtonian mechanics. What is new in Coriolis's approach is the extensive use of the concept of work with its mathematical formalism, which is quite different from Newtonian concepts. The new approach that he adopted can be better understood if we look at the two general approaches used for solving a given mechanical problem. One is to consider the variations of motion as the result of forces, in a Newtonian way. Alternatively, knowing the amount of work generated by the system, to try to find the "living forces" involved. In other words, we can use Newton's second law or work-kinetic energy principle. In 1844, 1 year after Coriolis's death, his master work was published with the title *Treatise on mechanics of solid bodies*. The book is divided into two parts. The first part, entitled *Mechanics of solid bodies and considerations about frictions*, is divided into three chapters as follows:

- Chapter 1: Notions about the velocity, the force, the weight, the mass, and the motion of a material point.
- Chapter 2: On solid body motion.
- Chapter 3: General considerations on the machines used to transmit work from the motor.

The second part is titled *Du calcul de l'effet des machines*, in which we focus on our main concern, which is the true development of applied mechanics, as we have seen after Lazare Carnot's work. It is also divided into three chapters. The book contains 367 pages with a structure similar to *Principes fondamentaux*. It begins with a review of the fundamental mechanical principles and concepts, in order to apply this knowledge to machines.

In this Coriolis text, the concept of work was used as an economic measure for comparing similar machines from the point of view of efficiency. From the beginning of his textbook, the entity of mechanical work was the most appropriate and suitable tool for evaluating machine dynamics performances.

On the first pages of his book, Coriolis states: “My purpose in this work is to highlights the questions about the economy of what is usually called force or mechanical power, as well as to provide the means to easily recognizes what are advantages or disadvantages of some machines constructions”. With respect to previous studies, he said: “It does not exist, if it does, I don’t know, about this subject, unless the works of Carnot and Guényveau; but at the same time, Pétit has introduced in Physical Proceedings, one concise memoir about the use of living forces principle, and after that Navier has published his additions on “Hydraulic Architecture” of Bélidor”. Coriolis also refers to Poncelet’s applied mechanics course in Metz and emphasizes the new denomination of work: “I have used in the present study some new denominations: I call by the name of work the quantity which is usually called mechanical power, quantity of action or dynamical effect, and I propose the name of dynamode to the unity of this quantity. I permit to myself a small innovation which is to call living force as the product of the weight to the height. This living force is the half of the product which is designated until now by this name, i.e., the mass times the velocity squared”.

The two citations above explain two important facts. Firstly, Coriolis was aware of the importance of Lazare Carnot’s achievements in treating the living forces problem besides tracing a trajectory to be followed in order to understand the history of the work concept, which is our basic concern. Secondly, we see the adoption of the term *work*, thus replacing the previous designations, as well as the introduction of the constant ½ to correctly quantify the living force. Some authors incorrectly attributed to Poncelet the use and introduction of the term *work* in mechanics.[17]

In the first chapter, Coriolis reviews the fundamental mechanical concepts. He uses the same definition of force as Newton’s: “Each cause which tends to modify the body motion, or to move it if it is at rest, is what we call force”. This statement is exactly what appears in Newton’s *Principia* to enunciate the second law. Not only in this citation, but also in the examples used, Coriolis’ concept of force differs fundamentally from d’Alembert and the Cartesians’ concept of force. Coriolis considers our primary idea of force experienced by us to support the action of a heavy body or by pushing it in some direction such that a certain velocity is impressed.

The review that Coriolis makes in the book encompasses practically the whole of rational mechanics. The concept of work in a formalized version appears for the first time on page 35, in the item entitled: “transmission of work principle in a material point motion”, which today is denominated as the theorem of work and kinetic energy. It is also in this point of the text where he introduces several definitions, such as moving force, resistant force, depending on whether the force decreases or increases the velocity of the body. As shown previously, these definitions had already been used by Lazare Carnot. Coriolis also defines quantity of work as an integral of the tangential component of force multiplied by the arc element of the trajectory of the particle, driven work and resistant work.

[17] According to the discussion in the footnote of the previous page, the utilization of the term work in technical literature is due to Coriolis, in spite of some authors such as René Taton having attributed this fact to Poncelet.

He enunciates what was denominated in the living forces equation, now referred to by him as the transmission of work equation, in the following form: "During any motion, the difference between driven and resistant work, due to applied forces to a material point, is equal to the increasing into the o particle living force during this time". This is the theorem of work-energy in a modern form.[18]

Before applying the living forces principle to machines, Coriolis briefly defines a machine: "In general we designate a machine a system of solid bodies in contact, in order to transmit the work of forces". With respect to the application of the principle of machines, he said: "The lost work in friction between two bodies belonging to a machine, will be calculate by an integral, during the motion under consideration, and then the element will be the product of the total friction to all contact points which have equal and parallel velocities, multiplied by length element of sliding of each body on the other". He observes that this calculation can be done only through experiments and explains in general how to set up this experience.

Another problem described by him and already studied by Carnot is the bodies' shock, with the purpose of applying to any machine in motion. As usually happens, the d'Alembert principle is used, now leading to Carnot's theorem. Coriolis uses the d'Alembert principle, but associated with the virtual works principle.[19] The enunciate proposed by him appears in its more general form, including for non-elastic bodies, as follows: "The difference between the living force due to velocities before the shock and that appearing by the medium velocities after the shock, is equal to the addition of two terms: (1) the living force due to lost velocities or gained by the shock effect, that is, the velocities which combined with that which appear after the shock, will give as result that which appear before the shock; (2) the addition of the products of the quantities of motion due to frictions during the shock, by the relative velocities from friction after the shock".[20]

To finalize these remarks about the work transmission principle, in Coriolis's language, he generalizes this principle to a machine, specifying all terms of the equation:

$$T_{\mathrm{m}} = T'_{\mathrm{m}} + T_{\mathrm{f}} + T'_{\mathrm{f}} + T_{\mathrm{c}} + \sum mv^2 - \sum mv_0^2$$

where:

T_{m} = Motor work (driven);
T'_{m} = Motor work transmitted by machine;
T_{f} = Work lost through friction in external parts of the machine;

[18] It is important to note that, in the nomenclature as well as in the concepts, the metaphysical aspects that followed up until Carnot progressively disappear.

[19] It is the modern form for applying the d'Alembert principle. When we look at the velocities decomposition proposed by him, the velocities which are destroyed in shocks are seen separately as an equilibrium problem.

[20] As in previous situations, an energy balance shows two sources of energy loss, shocks and frictions. This is clear with the equation presented in last paragraph.

T'_f = Work lost through friction in internal parts of the machine;
T_c = Work lost from shocks;
V_0 = Initial velocity;
V = Regime velocity.

It is in the final Chapter 3, dedicated to a review of rational mechanics, that Coriolis introduces his economic considerations and again defines what a machine is: "A machine in its usual meaning, is an assembly of bodies in motion, whose configuration is to build a kind of chanel through which work like a fluid can be transmitted, the most efficiently possible, to the points where they are needed. It is lost from time to time due to frictions or the deformation of bodies, or it will spread out until disappear".[21]

It is important to notice the difference in conceptions about a machine between Carnot and Coriolis. The first sees a machine as a system for communicating motion, as something that propagates in a chain. Coriolis has a vision of a machine as a hydraulic device where the work is associated with a fluid in motion and, as it moves along its path, the fluid spreads and getting looser because of resistance or other kinds of loss until it becomes insensible. These considerations have remarkable epistemological importance to the history of science. The images or analogies did not allow for the realization of the loss of energy and the transformation of a certain amount into heat. Only some years later was the idea that energy can be degraded within machines scientifically accepted.[22] Coriolis's idea about work is clear in this statement: "We cannot produce nothing necessary to our needs unless by displacing bodies or changing their shapes; what we cannot to do in earth's surface unless displacing resistances, and acting efforts in the sense of motion. Then, an usefull thing that is the capacity to produce thus the displacement of a force in the same sense, that is, the capacity to produce what we call work".

Therefore, Coriolis supports his economic analysis in a conception of production similar to Navier. According to him, all kinds of production can be represented by a mechanical action which consists of opposing resistances to bodies, being displacements or deformations. In other words, work can appear only in these two types of action. Coriolis, as other polytechnician engineers did, uses economic concepts that are common in this period.[23]

Coriolis also ascertains that the capacity to produce work is limited in time being relative to a given place, and he concludes that this capacity is merchandises,

---

[21] The image used by Coriolis to visualize a machine is interesting. Work is communicated to the machine pieces among them as a kind of *vital flow* which feeds the machine and puts it in motion. The losses due to resistance, friction or deformations are seen as holes through which the flow moves.

[22] In spite of the work degradation inside the machines and part of it being transformed into heat, the development from mechanical work to other energy forms was necessary for a new synthesis in physics, made by the discovery of the first law of thermodynamics.

[23] Naturally, in Coriolis´s time, it was common to associate economic production with mechanical action, because this is what was done in production, meaning the replacement of labour (human work) by machines. In the text, we see a synthesis of some basic concepts used by Jean-Baptiste Say who had significantly influenced the polytechnician engineers.

in the same form used by economists, as good, useful and rare. Thus, he states: "Independently of how the motions, through the animals, the water or the air in motion, the carbon combustion, the fall of bodies, it is limited in each time, in each place; it is not created by the willing. What the machines do is to employ and to save the work without to increase it; since that the capacity to produce is sold, it is possible to buy it, and we save it how the other things that are not in extreme plentiful".

In addition to this consideration made by Coriolis through stating that merchandise is not the work but the capacity to produce it, the citation above highlights that, from the physical point of view, the idea of energy does exist, appearing as work, for any type of motor that produces it, whether a man's force, an animal's force, the force of air, the force of fire or any type that we can imagine. By making this distinction between work and the capacity to produce it, Coriolis made an important attempt to eliminate the mistake found in Navier's work when he analyzed the product and its expenditure. This distinction is also found in Marx when he established the difference between work and its potentiality, that is, the labour force. This question studied by Coriolis provides an analogy between their theoretical constructions and that proposed by Marx in physical work and abstract labour, respectively.[24]

This analogy remains, with respect to the equivalence norm already established by both thinkers. To Coriolis, it is the transmission of work principle which provides this norm within the heterogeneity of the concrete processes in the production and the existence of machines which permits the practical establishment of the measurement of equivalence. Coriolis affirms: "If the machines do not exist to our utilization, two different displacements would be two things of different nature which do not admit any mathematical basis for its evaluation: these displacements as useful things whose values cannot be established using mathematical calculations. Although, the machines provide the way to put the displacements in similar basis and this is what we have to similar quantities".

The transmission of work principle, which has a central position in Coriolis' work, is based on rational mechanics being later reformulated to include economic aspects and thus permitting him to come back to Navier's physical–economic elaboration. He begins by assuming that work is conserved during machine working and concludes that there are no losses. Unlike Navier, Coriolis applied the above-mentioned principle to postulate a theoretical model with the purpose of calculating the production and the expenses.

In order to apply the transmission of work principle more easily, Coriolis made some simplified assumptions, like the machine efficiency being equal to one. This means that the resistant work is equal to the useful effect. If we apply this hypothesis

[24] Although we can trace a parallel between labour (human work) and that made by a machine, the differences are significant and it is exactly those differences which limit the use of a theoretical model from physics to explain economic phenomena. In a capitalist society, labour is merchandise, but a very singular merchandise, because it is unique in creating value. However, what the capitalist buys is not the labour but its utilization.

to production, the useful effect, thought of in terms of work, is proportional to any physical measurement used to measure production, such as volume, weight, length, etc. In other words, in the idealization made by Coriolis, with machine efficiency equal to one and machine motion being uniform motion, useful work is equal to resistant work which is also equal to driven work. Thus, the principle of work transmission provides the common measurement in the production context.

Starting from the hypothesis that work is the search and satisfactory quantity in the economic production context in order to compare motors, it is sufficient for it to be admitted that a machine can transform motor work (driven) into some amount of useful work. In Coriolis' words: "If we wish to compare two capacities of producing motion, it will be sufficient to conceive that machines were built to help us to apply these capacities to fabrication, for instance wheat milling... Of course the two milling values will be measured by the number of litres of wheat milled; but these last are proportional to the quantities of work motor produced in each machine, and we conclude that the two capacities of motion will be values proportional to the quantities of work that can be produced in these machines".

It is clear by the above citation how Coriolis, using the concept of work provided by rational mechanics, is able to use that concept as an economic comparison tool between motors and arrive at the same conclusions as Navier, but in a entirely new context. The motion accomplished by the two thinkers is quite different. To Navier, work was, from the beginning, one economic quantity and one *mechanical currency*, as we have seen, and starting from this stage, he has shown how the work could be physically measured. The theoretical change made by Coriolis is that he emphasizes how a physical concept, the work concept, allows for a comparison of an economic nature to be done, even in an approximate form.[25]

Another important problem studied by Coriolis is the introduction of time for undertaking certain work. From the mechanical point of view, the capacity to undertake work in an interval of time is measured by the mechanical power the definition of which is the work done by some agent in the unity of time. Coriolis's text has some ambiguities. Initially, he recognizes that time is a criterium to compare motors: "Two similar displacements, like the transportation of two weights, made in different times, are two useful things of different nature, which under the time relation, do not admit geometrical comparison". Later in the same text, he uses the same arguments already used by Navier: "We should remark, however, when we should operate a machine some quantity of similar displacements, and the cost involved is the same in general, to operate simultaneously, but successively, we can introduce the time as an element of value of these operated quantities. Suppose that, for instance, we try to employ ten men to lift weights: if we wish operate this kind of elevation, we could always do it by employing twenty men; without no more cost, the same effect will be done in a medium time by the half. This decreasing in time can thus be obtained as we wish, and cannot be paid".

[25] In Coriolis' works. the metamorphosis of the concept of work towards an economic field is more remarkable than in that of other polytechnician engineers, which we hope was made clear in the text.

Coriolis' analysis is quite similar to Navier's, as are his considerations on *mechanical currency*, which work as a quantity independent of time. In other words, the time consumed in the accomplishment of certain operations, a machine, for instance, or a certain task, for some human labour, does not make any difference in terms of value. This supposes that work is time-free and the increase or decrease in time does not vary the cost involved.

Obviously, this analysis made concerning the time is not correct because it supposes that time is an infinite resource being associated only to use-value but not to exchange-value. Although Coriolis' analysis does not survive even the common sense arguments, the question of Coriolis' mistake remains. An attempt to explain this fact may lay in the fact that, in the physical definition of work, the time does not appear to be creating a contradiction with its main purpose, which is to legitimate the concept of work economically as the measurement of value. However, in the last chapter of his book, he comes again to this question and changes his opinion with a contrary argument: "We have repeated what was said in the first chapter, that work, being the main element from what we pay to motion, and the unique belonging to the domain of the exact measurements, is not the only belonging to the motion value. Then, the volume, even that is the main element to the value of several useful components, is not the unique to be considered to attribute values".

With the above citation, Coriolis changes his opinion with respect to the beginning of the book and again discusses the questions of an economic nature, presenting a critique of the Navier physical-economic model. His remarks are concentrated on the common measurement tool for the net product (useful effect) as well as for the expense (total work). He is looking for an optimization of machine efficiency, following his predecessors. This can be obtained, in terms of work, by the ratio between useful work and total work. In this theoretical model, it is implicit that part of this useful effect, in the economic sense, can be measured by work since it can be distinguished from the lost work. On the other hand, the energetic efficiency is the main concern and the priority of entrepreneurs. By means of these considerations, we should take care from the economic point of view about the generalization of this analysis. He states: "We have to take into account, within the economy of work, the third part of the machines, i.e., what immediately operates the useful effect… what is that? The mechanical effects of machines are: (1) the lifting of weights; (2) the breaking or changing in bodies' shape; (3) the frictions to overcome to operate the slowly displacement of bodies; (4) in the fast transportation, which is the velocity production".

Of course, the four types of useful-effects mentioned above give rise to the most complex problems. The first three of them could be analized by Navier's model. However, certain questions, for instance, how to distinguish without ambiguity what is useful work or pure loss, acquire a relative character due to the complexity which can involve a machine motion. Thus, it remains economically arbitrary to speak about lost work.

And yet, concerning the effects to which we referred in the last citation, Coriolis states: "The first three effects absorb completely a certain amount of work, which cannot appear again, at least by the moment. Thus, the bodies broken or deformed, the frictions overcome, the lift bodies, since that do not repeat this operation, are consumed quantities of work which cannot be transmitted. This quantity will be responsible theoretically by producing these effects; but is necessary always to consume a less quantity, because the communicated velocities and the modifications which occur in the bodies around...".

We can go back to the question of mechanical efficiency and the impossibility of a machine efficiency being equal to one. This is explained by thermodynamics. Otherwise, it is possible to think only from a mechanical point of view. The mechanical efficiency equal to one would mean an ideal machine, which would be economically absurd or even impossible. From the mechanical design point of view, this kind of machine can transmit motions without shocks and also with infinitely small variations of velocity. Such a type of machine is neither viable nor desirable.

What is practically viable is to maximize the mechanical efficiency through continuous and pragmatic searching, by considering a machine as a black box consuming a certain quantity of work, where it is perfectly possible to distinguish useful effect and expenses. Based on this type of analysis, the objective of the entrepreneur is sometimes an empirical task. He must reduce the expenses which clearly appear and came from the living forces, without achieving the absolute maximization.[26]

Within this technical-scientific framework, Coriolis is aware of the limits imposed by the physical-economic project of polytechnician engineers founded on the knowledge that came from the mechanics of the eighteenth century. When the concepts of rational mechanics are used to solve the economic problems of machines, they are not sufficient to accomplish them. Thus, a new machine science becomes necessary: thermodynamics, as we know, founded a few years before by Sadi Carnot .[27] Sadi, Lazare Carnot's son, thought the transformation of living force in work to be an irreversible process, an energetic transformation which, even theoretically, involves losses. To achieve this point of development, it was necessary to abandon the rational mechanics framework, consider a more broad physical process and create a new currency for conversion of the physical processes into energy.

The importance of the history of the work concept associated with the science of polytechnician engineers is to show the genesis of thermodynamics, as is the end of the possibilities of rational mechanics to explain thoroughly all the machine phenomena.

In the analysis of the process of incorporation of the work concept to economic thought, it is fundamental to take into account the schools of economic thought in that period as well as the socioeconomic transformations. Within the field of

[26] It is possible to realize clearly the difficulties of the theoretical model and the need for a new one, which will be provided only by thermodynamics.

[27] This is the way thermodynamics arose, initially as a machine science and later as a more complex science of the general nature processes.

mechanics of machines, it was necessary to develop a science specific to them. As we have shown, this begins with Lazare Carnot and continues with the elaborations of polytechnician engineers using the concept of work in all theoretical constructions, including those of Lazare Carnot. Further developments occur with the rise of thermodynamics, which provides the conceptual framework necessary to study the machines, even if we need to study them from a kinematic point of view. Obviously, for this particular case, motion is partially transformed into heat with thermodynamics being the suitable science to be used. For the purposes of our study, we should analyse the developments which led to industrial mechanics as a consequence of the development of machine mechanics.

7.5 Poncelet's Contribution to Applied and Industrial Mechanics

Jean-Victor Poncelet was born in Metz on 1 July 1788, the son of a rich land owner and lawyer. In 1807, he entered the Polytechnic School, where his teachers included Monge, Lacroix (1765–1843), Ampère (1775–1836), Poinsot (1777–1859) and Hachette (1769–1834). He graduated at 22 in 1810. He joined the Engineering Corps and went to Metz to study at the Ecole d'Application. After 2 years of study, he graduated, having reached the rank of Lieutenant.

In June 1812, he was called to take part in Napoleon's Russian campaign. He was arrested and held in prison from March 1813 to June 1814, when he returned to France. During this imprisonment, he recalled the fundamental principles of geometry. He thus began to develop projective properties of conics. From 1815 to 1825, he was Captain of Engineers at Metz, working on the construction of machinery in the arsenal at Metz and teaching mechanics in the military college. Using the work of Navier and Coriolis,[28] as well as his own discoveries, he gave the first course on *Applied Mechanics to Machines*, which would be transformed into a book in 1826. In 1829, Poncelet published the Course of Industrial Mechanics, which, after his death, became an important reference and fundamental textbook.

Poncelet dedicated a large part of his professional life to the teaching of engineering, improving his courses on theoretical and applied mechanics. Machines were also always the reason for his concerns. In 1825, he made a tour of Europe to study the machines used in various countries. He visited Germany, Belgium and France itself, leading him to rethink the application of machines to industrial processes (Fig. 7.7).

---

[28] As remarked previously, Coulomb and Lazare Carnot are the true precursors of the polytechnician engineers. Coulomb created a new science of friction and Carnot developed the first general theory of machines. Coriolis systematized and developed Carnot's theory within the framewok of rational mechanics.

Fig. 7.7 **a** Jean-Victor Poncelet (1788–1867), **b** Introduction à la Mécanique Industrièlle

In order to keep himself up to date on the new types of machine that were being developed, he took part as one of the organizers of the large fairs of London in 1851 and Paris in 1855.

Poncelet received many honors in addition to being elected to the Academy of Sciences. He was an officer of the Legion of Honor and Chevalier of the Prussian Order. Many academies and scientific societies elected him to membership, including the Royal Society of London, the Berlin Academy of Science, the Imperial Academy of Sciences of St. Petersburg and the Academy of Science of Turin.

After a long and painful illness, Poncelet died in Paris on 22 December 1867, at the age of 79. In the following year, the Prix Poncelet was endowed by his wife to carry out Poncelet's dying wish that the sciences be advanced. The prize was awarded for work in pure mathematics or mechanics by the Academy of Sciences starting from 1876.

The most important characteristics of Poncelet's work can be summarized as follows:

Poncelet—Polythecnicien. The project of the experimental and physical mechanics, which is a restricted version of the Laplacian project, mobilized a group of French *polytechniciens* during the first-half of the nineteenth century, including Poisson, Navier, Coriolis, Cauchy (1789–1857), Saint-Venant (1797–1886) and others. The most representative name in the field of the new mechanics is obviously Poncelet. It is possible to separate many aspects of Poncelet as a *polytechnicien*:

- *Inventor.* A model of a bridge presented to the Fortification Committee in 1820, constructed and much used; the hydraulic wheels with his name, the design of which was published in1827 with the title *Mémoire sur les roues hydrauliques a aubes courbes*[29]; a dynamometric device to measure the work of motors and machines; an apparatus to discover the laws of non-uniform motion of machines experimentally.

- *Theoretician of the new physical mechanics.* He proposed a geometric representation of the molecular constitution of matter which can explain macroscopic phenomena such as elasticity and deformation of bodies.
- *Experimentalist.* He idealized and performed several research programs with an experimental nature about water flow through orifices;
- *Lecturer.* Working on theory and experiments, Poncelet developed a modern theory of machines. Applying this knowledge in his courses, he made this knowledge known to engineers, students, industrialists, craftsmen and workers.

Poncelet—Theoretician. Facing the new problems presented by 'physical mechanics', Poncelet had to formulate a theory that could explain the constitution of matter and allow him to make conclusions about the behavior of natural bodies. Poncelet was one of the few theoreticians who adopted the concept of bodies as systems of molecules maintaining small distances between each other. These systems are subject to attractive and repulsive forces, similar to the general proposition of the Laplacian project. In addition to these assumptions, he adopted a geometric representation in which he discussed the problems which appear in macroscopic scale.

On the other hand, Poncelet postulated a philosophical point of view based on a scientific realism, which analyzed the natural world using models. This means that he postulated the existence of theoretical entities that were not directly observable but which were nonetheless as real as the physical world.

In relation to the general model of the molecular constitution of matter, Poncelet proposed some specifications. Similar to Poisson, Poncelet also proposed that action between two molecules, being attractive or repulsive, depends upon the respective distances, and that a kind of law could be derived. With the expansion of distance, the forces decrease as also prescribed by the Laplacian project.

Poncelet was the first to propose a graphic representation of the variation of the intensity of forces as a function of the distance that separates two molecules. Based on this model, he explained macroscopic phenomena as elasticity. The difference between different materials can be explained by different geometric curves representing their molecular action.

Poncelet—Experimentalist. Working with the military engineer Joseph-Aymé Lesbros (1790–1860), Poncelet carried out a series of experiments trying to understand the laws of water flow through orifices. Based on numerical findings, the final report of this investigation is a good manual for experimentalists. With its help, it is possible to reproduce and perform the same experiments. In addition, they proposed a kind of experimental method within this field of investigation.

---

[29] In the introduction to Poncelet (1827), we find a historical summary of Poncelet's works related to water wheels. The first of these memoirs won a prize in mechanics at the Royal Academy of Sciences in 1825. It was also published in Physics and Chemical Proccedings in 1825 and 1826. A second memoir appears in 1826 where the achievements of several experiences with a new water wheel are presented.

Also common in Poncelet is an experimental procedure that is a kind of 'mental experience', in other words, a mental construct from possible experimental situations. In an example of a similar situation, in 1798, Lagrange used an imagined pulley system to prove the virtual velocities principle which belonged to basic rational mechanics.

In relation to the experiments in the applied mechanics field, it is important to explain that this operation cannot be reduced to a production activity. In order to produce a set of valuable results by means of experiments, it is necessary to have experimentation protocols which establish the conditions of findings. On the other hand, it is worth emphasizing that the experimental works carried out by eighteenth century engineers were neither sufficient nor rigorous enough to create valuable scientific knowledge.

Poncelet's Applied Mechanics Courses. Poncelet participated significantly in the spreading of mechanical knowledge in different forms. Probably the most important form was his teaching of applied mechanics courses. From 1825 to 1834, Poncelet gave a course on machines in the Metz School of Artillery and Engineering to engineers who had finished the Polytechnic School and intended to follow a military career. He also gave another course with the title of industrial mechanics to workers in Metz from 1827 to 1830.[30]

In 1874, François Xavier Kretz published a book that is part of Poncelet's lessons in École d'Application de l'Artillerie et du Génie de Metz. The course, "Cours de mécanique appliqué aux machines", grew out of lectures that Poncelet had given from the above-mentioned period. The course follows the general topics presented below:

7.6 Applied Mechanics to Machines

Section I. General considerations about machines in motion.
Section II. Main methods for controlling the action of forces applied to machines and how to transmit velocities with given ratios.
Section III. How to calculate the resistances of pieces in uniform motion.
Section IV. On the influence of velocities variations over the resistances.

Some years later, this course was complemented with two more sections, including practical hydraulics, gas motion, water wheels and steam machines. In Section I, Poncelet had the objective of explaining the general principles applied to the theory of machines in motion and how the mechanical properties can prescribe the conditions to be satisfied by machines. He also established the general

[30] The history of Poncelet´s courses on applied mechanics begins in 1825, when he was admitted to the Metz School (Poncelet 1874). In 1827, Poncelet founded a professional course of public character at Metz, tax free and addressed to workers, entitled: Afternoon Lessons on Industrial Mechanics. In 1838, he was invited to Paris to create the Course of Mechanical Physics and Experimental. These are the most important points concerning Poncelet´s career as a lecturer.

equations governing machines in motion by applying the living forces principle. In Section II, Poncelet developed the study concerning the systems of regulation of the action of forces in order to assure uniformity of motion. This section was dropped after 1826.

In Section III, different kinds of friction were studied based on Coulomb's findings, which were taken into account by Navier and Belidor. In addition, Poncelet also used the approaches indicated by Lazare Carnot in his "Principes" regarding motion with high changes of velocities.

The Sorbonne course given from 1838 to 1848 represents the final period of his life as a professor. The title of this course was Applied and Experimental Mechanics. Poncelet used elementary mathematics such as geometry, arithmetic, elementary algebra, trigonometry and logarithms. This course was heavily illustrated and rich in descriptions of measurement instruments and similar apparati, with detailed instructions for their use. The program reproduced below refers to a course at the Sorbonne in 1840–1841.

7.7 Physical and Experimental Mechanics

Part I. Summary of fundamental notions related to the constitution and physical properties of bodies.

Section I. On motion from a geometric point of view.
Chapter I. Fundamental preliminary notions.
Chapter II. The laws of motion.
Chapter III. Geometric transformations of motion.
Chapter IV. Composition and decomposition of absolute and relative motion.
Section II. On forces and the abstraction of motion that they produce or destroy.
Section III. On the resistance of solids.
Section IV. On the communication of motion by forces.

The above topics are only the primary ones, because the sections are further divided into chapters and the chapters into lessons. Figures shown below are illustrations presented in that course (Figs. 7.8 and 7.9).

Some remarks need to be made about Poncelet's course in Sorbonne. Part I, dedicated to summarizing 'Fundamental notions and physical properties of bodies', is similar to the corresponding part of his 1841 'Introduction to Industrial Mechanics.' In relation to Section I on 'Motion considered from the geometric point of view', this was the first course on theoretical kinematics either in France or elsewhere in the world.[31] In 1834, Ampère suggested the creation of a science of motion abstracting the conditions of its production, in other words the study of

[31] Teun Koetsier, in his paper *The Case of Kinematics, The Genesis of a Discipline*, published in the Proceedings of HMM 2012, Amsterdam, May 2012, wrote: At the end of 1820s, Ampère coined the word kinematics in his *Essai sur la philosophie des sciences*, or *Exposition analytique d'une classification naturelle de toutes les connaissances humaines*, Paris, vol. ½ (1834–1843).

Fig. 7.8 Dynamometric brake

motion without any consideration of forces. After Poncelet, this new science was definitively incorporated into mechanics.

Returning to the course program, after Section I, Poncelet analyzed statics which can be visualized from the point of view of theoretical mechanics as well as the effects of forces implied in motor studies that are part of applied mechanics. These two approaches are unified by a systematic utilization of the concept of work. In addition to the study of the theoretical aspects of work, Poncelet developed methods and instruments to measure this.

In relation Section III, on the 'Resistance of Bodies', this is actually the same as strength of materials.[32] We can find tensile, compression and shear forces, bending and torsion applied to prismatic sections. The course is of an advanced level and proposed several new concepts and methods at that time.

[32] It is easy to realize the amplitude of this course in terms of the contents presented from the laws of motion to the new notions of strength of materials.

Fig. 7.9 An alternative motion machine

Poncelet also referred to a course given in 1837–1838 by Saint–Venant in *École des Ponts et Chaussées* entitled Mechanical Lessons. It was the first course on the strength of materials that used the new concept of the theory of elasticity.[33]

Around the year 1830, industrial mechanics was a flourishing movement, with a fundamental contribution to the science of engineering, as well as the contributions of European machines constructors since the end of the Middle Ages. In France, this practical mechanics had been developed in the context of an original socioeconomic framework with a military corps of engineers organized for the defense of country and thus creating some demands in terms of engineering knowledge as fortification, for instance. This was the general characteristic for the rise of industrial mechanics.

At the beginning of the nineteenth century, a kind of turning point appeared. It was the influence of political economics on the new science of machines. This period is generally described as the Industrial Revolution, first in England more or less around the first decades of the nineteenth century, and spreading to other European countries and the United States. The birth of political economics had an

[33] The main contribution of Saint-Venant was in the field of the mathematical theory of elasticity, but he also developed much in the elementary strength of materials, especially with the theory of bending of bars. As described by Timoshenko in his *History of Strength of Materials*, Saint-Venant was the first to examine the accuracy of the fundamental assumptions regarding bending.

important effect and caused a profound rupture in the creeds concerning the relation between the State and Society. The society was no longer thought of according to a mercantilist conception as a tool in favour of state power, but, on the contrary, that the power of the state should be redirected in favour of the society and its economic development. Obviously, these ideas influenced the polytechnician engineers of this period.

The questions of economic nature appear more frequently in Part IV of Section I of the course *Applied Mechanics to Machines*, entitled *Establishment of Industrial Machines*. At the beginning of this item, referring to general questions, we can read: "We already said that the essential conditions of a consistent establishment that provides the maximum useful effect or the maximum of work undertaken is a minimum of expense in terms of driven work and money, such that the unity of work of each specie be provided by the least possible price. In order to study this question in all generality, it would be necessary to vary all the data that it depends, in the relations that are related to both the useful effect and the expenses; but making abstraction of the price in money, which varies according to the times and localities, we cannot study thus the question of machines establishment." He continues: "We are satisfied in make the decomposition in several different accounts for a separate treatment: thus we study successively the action of motors on the receptors, the tools or operators over the matter to work, to displace, etc., and thus we do not go to material pieces which are used only to communicate motion".

This citation itself explains Poncelet's strategy related to *Industrial Mechanics*. He understands a machine as a complex chain to communicate motion, as well as the work being a kind of mechanical currency as Navier had proposed; the question of optimization of a machine's operation conditions for less cost is a complex problem.[34] His approach is then to tackle the global problem by means of decomposition in parts and taking into account that the lost work in passive resistances and shocks in transmission and communication elements of motion is small and that it should concentrate his analysis in extreme elements of the motion chain, that is, in the operator (the driven motor) and receptor (device which performs work). Poncelet continues his analysis discussing a series of conditions and requirements that must have the driven motor and the operator in order to obtain a more favorable useful effect.

---

[34] In his second memoir on water wheels, Poncelet optimized the speed which water should have in order for its effect on the wheel to be at a maximum. Considering V the water speed on the wheel, H the height from where water falls, m the mass of water flowing during one second, g gravity acceleration and v the constant speed acquired by wheel. $V - v$ will be the relative speed with which water will go up, $(V - v)/2g$ will be the height raised. Thus, $(V - v) - v = v - 2g$ will be the absolute water speed leaving the hydraulic wheel. Because this speed will be zero in order to obtain the maximum effect, we have $V - 2v = 0$ or $v = ½ V$. This means that the water wheel will acquire half the flowing speed as well, as predicted by the theory of hydraulics with common blades.

In Section II, Poncelet analyzes the primary means of regulating the action of forces in the machines and of transmitting the velocities in given relations. His purpose is the optimization of machine operation. In Section III, his main concern is to calculate the passive resistances in pieces in motion which influence the calculation of machine expenses. Thus, several kinds of relative displaments among parts of a simple machine are studied, in a similar approach to Coulomb's friction studies.

In the last section, the basic concern is the *influence of velocity variations on the resistances*. As remarked previously, Poncelet uses a classifying scheme for machine motion similar to Carnot and his study on passive resistances closely follows this scheme, starting from the calculation of static friction, moving on to the case of friction of pieces in uniform motion, and finally to the cases where the velocity varies from an alternative motion up to more complex variations. The cases of loss through shocks among pieces in motion are studied using the loss of living force, as per Lazare Carnot. However, Poncelet's approach is quite similar to Coriolis' method, with the same mathematical tools and modern formalization, as well as the same kind of decomposition of machine problems with the purpose of optimizing all elements and processes with significative influence on machine expenses.[35]

Finally, the participation of Poncelet in the process of incorporation of the concept of work into economic thought is significant. The work concept with economic aspects or the physical-economic concept appears when the practical mechanics find rational mechanics at the moment when the mechanics formalization can be used by a kind of machine economics. This new social and economic framework emerged in Europe, mainly in France and England. Therefore, this period was contemporary to the birth of classical political economics.[36] In fact, it is exactly this scientific discipline that will provide the new elements and concepts which made possible this passage or enrichment of a concept from physics acquiring other dimensions coming from economics. In this context of great development of mechanics with new economic ideas, the concept of work operating as a bridge-concept between the two sciences is transformed and used by engineers and economists. Poncelet's work is part of this process by organizing a new form of presenting mechanical problems, incorporating and developing a new science of materials, either solids or fluids, with a fundamental role of the concept of work.

---

[35] We are referring to a kind of decomposition of the text into *machine elements* as is usually used now by applied mechanics or mechanical design, such as shafts, bearings, belts, pulleys, etc.

[36] Classical economics or the classical school of economics belonged to Adam Smith, Malthus, David Ricardo and John Stuart Mill. Smith inaugurated the classical period with *Wealth of Nations*, published in 1776. Overcoming the physiocratic school (Say 1983), Smith adopted as a central theme productive activity and work as a source of richness. As is known, the physiocrats had an exaggerated agrarian conception.

References

Coriolis GG (1829) Du Calcul de l'Éffet des Machines. Carilian-Goeury, Librairie, Paris
Coriolis GG (1832) Sur le Principe des Forces Vives dans le Mouvements Relatifs des Machines. Journal de l'École Polytechnique, vingt-unième cahier, tome XIII
Coriolis GG (1990) Théorie Mathématique des Éffets du Jeu de Billiard. Éditions Jacques Gabay, Paris
Coulomb CA (2002) Théorie des Machines Simples. Librairie Scientifique et Technique Albert Blanchard, Paris
Galileo AF (1987) Descartes e o Mecanicismo. Gradiva, Lisbon
Koetsier K (2003) A Note on Gaspard-Gustave de Coriolis (1792–1843). In: Proceedings of the 11th world congress in mechanism and machine science, Tianjin, China
Merlet JP (2004) A note on the history of trigonometric functions and substitutions. Proceedings of HMM 2004. Kluwer Academic Publishers, Dordrecht
Oliveira ARE (2004) The contribution of coulomb to applied mechanics. Proceedings HMM 2004 Symposium. Kluwer Academic Publishers, Dordrecht, pp 217–226
Poncelet JV (1827) Mémoire sur les Roues Hydrauliques a Aubes Courbes. Secong Mémoire, Librairie de Me. Ve. Thiel, Place St. Jacques, No. 4, Metz
Poncelet JV (1870) Introduction a la Mécanique Industrielle, troisième edn. Gauthier-Villars, Imprimeur-Librairie, Paris
Poncelet JV (1874) Notice sur l'Introduction à la Mécanique Industrielle et sur le Cours de Mécanique Appliquée aux Machines. Gauthier-Villars, Paris
Say JB (1983) Tratado de Economia Política. Coleção Os Economistas, Editor Victor Civita, S. Paulo

Chapter 8
Conclusions

If on the one hand machinery has the tendency constantly to throw workers out, whether from the mechanical workshop itself, or from the handicraft enterprise, on the other hand it has the tendency constantly to attract them, since once a particular stage of development of productive forces is given, surplus value can only be increased by increasing the number of workers employed at the same time. This attraction and repulsion is the characteristic feature of machinery, hence the constant fluctuation in the worker's existence.

(Marx 1861-1863 Manuscripts. Fragment of "Relative Surplus—Accumulation")

In this chapter, by way of a conclusion, after some general considerations about the investigation, we will look in a synthetic manner at three questions that are closely related to the central theme being dealt with here.

Initially, we would like to note that although our investigation has almost entirely been concerned with the history of science, it should be seen in a more general context as a contribution to the ontology of labour. This involves placing it in a perspective in which labour as an activity, whether or not it adds value and irrespective of its economic repercussions, continues to be a fundamental research theme in both the natural and social sciences.

Our greater objective in presenting the reflections contained in this study is to fill in at least part of a significant gap—the almost complete absence of studies about the concept of labour as physical action, as a connection between man and nature without a greater concern that after a specific historical period this mediation became responsible for an economic value added to the product of labour called merchandise. Part I, called 'Conceptual Genesis,' is concerned with analyzing the origins of the physical concept of labour, its more general relations with the concepts of physics and with the history of rational mechanics from the point of view of its fundamental principles. Part II, entitled 'Instrumental Genesis,' looks at the developments and transformations of the concept after it emerged from

A. R. E. Oliveira, *A History of the Work Concept*, History of Mechanism and Machine Science 24, DOI: 10.1007/978-94-007-7705-7_8,

mechanics and was applied to machines, including the phase where it acquired economic characteristics. We will now look at some of the particularities of the two periods investigated.

The first begins with the birth of mechanics with Aristotle, then passes through a long period of gestation, going from Galileo to Newton, until the establishment of scientific method, reaching its apogee with Lagrange's formalization in the last decade of the eighteenth century, and ends in the second decade of the nineteenth, with the disappearance of the virtuous cycles of discussions about the principle of virtual work. The second period commences with the first application of the vis viva principle to machines by Daniel Bernoulli in 1738, which was developed into a general theory for machines by Lazare Carnot using Newton and d'Alembert's conceptual framework of mechanics, and ends in the third decade of the nineteenth century when the concept of work was applied to machines by polytechnic engineers using Carnot's theory.

There is a difference between these two periods which it is necessary to explore bettering more depth. While the first period coincides to a large part with the history of mechanics and stretches across centuries where there are multiple and varied contributions resulting from the creativity of genial figures in many different countries, this did not happen in the second period. In addition to being concentrated in time, it occurs almost entirely within the analytical tradition that emerged from rationalism in eighteenth century France. Let us examine this. After Daniel Bernoulli used the vis viva principle to study machines in a pioneering form, Charles Borda and Coulomb investigated hydraulic wheels using the same principle, as well as applying it to windmills. After this came Lazare Carnot and the generation of polytechnic engineers who succeeded him. All of these developments took place in France, and with the exception of Daniel Bernoulli who was Swiss, all involved were French. Poncelet, as we have seen, referred in a very superficial form to Smeaton as having defined the concept of mechanical power and to his unit, horse power (HP), which is still used nowadays to measure the capacity of an engine.

Lazare Carnot is one of the most interesting figures in eighteenth century French science. A high ranking military officer who held important positions in the republican government which followed the fall of the absolute monarchy, he made a singular and original contribution to the mathematics and mechanics of his time. His theory of machines had the great merit of presenting a solution for what we call the 'dilemma of rational mechanics,' and was the starting point which bloomed in the hands of the second generation of engineers who left the Polytechnic School, as well as being a fundamental contribution to engineering science. Furthermore, Carnot used a revalorization of maths, especially geometry, as a description of the laws of mechanics in the best Galilean sense, in a different manner and even to an extent opposed to Lagrange's eminently analytical and algebraic project, which dispensed with the use of figures. All of this was done using differential and integral calculus in a systematic and current form.

Historians of science are unanimous that from the point of view of the construction of a science of machines (theory) there are no intermediaries between Carnot and Coriolis. There are thus 26 years of interregnum. Some though point to the updating of Bélidor's 'Hydraulic Architecture' by Navier (1819) during this period as an important contribution to Coriolis' venture.

This particularity that the most important studies which used the physical concept of work for a science of machines were concentrated in France, and which were also part of the conceptual framework of rational mechanics, has to be seen in a broader context. Until the end of the 1820s a strong analytical traditional existed in French science which was reflected in the emergence of Fourier's mathematical theory of heat in Fourier (1822) in Chap. 3, as well as in the analytical studies of Ampère (1823) on electromagnetic propagation in 1823. In the following decade we can identify in England an analytic type investigation of the utmost importance for mechanics, which also followed the French tradition, this time as a reworking of Lagrange's mechanics through the legacy of Maupertuis' principle of minimum action. In 1834 Hamilton presented to the British Association his work "On the Application to Dynamics of a General Mathematical Method Previously Applied to Optics" (Crowe 1993). In the 1840s, George Gabriel Stokes' (1819–1903) mechanical theories on ether appeared, while in the 1850s Maxwell started work on his theories of electromagnetism, which would be completed in the 1870s. Hamilton also made a fundamental contribution to vectorial calculation, as well as achieving fame, with his creation of quaternion's in 1843.

In order to obtain a more realistic evaluation of the significance of the incorporation of a general theory of machines in rational mechanics we will return briefly to Galileo. From the more intuitive and qualitative point of view, the ideas of work and energy are very old. In relation to a conceptual approach taking mathematics as a description of the physical world, these ideas are much more recent. It was the research carried out by Galileo which orientated the science of motion and in it we can clearly identify two directions. Much before establishing a direct dependency between the velocity acquired by a body and the duration of its fall, Galileo first proposed that it was proportional to the distance covered. Realizing his error he returned to the first proposition that related velocity to the time and motion in free fall with time being squared.

Adopting the second path, Descartes proposed the current concept of the quantity of motion and following the same path Newton reached the concept of force. Nevertheless, Galileo's initial attempt to relate the velocity initially acquired to the distance covered would still be shown to be fertile. Galileo revised this question in light of his experiences with the pendulum, finding that the body reacquired the same velocity after returning to the initial altitude. Huygens and Leibniz expanded the discussions of this question and laid the foundations of what would later be called the principle of work and of vis viva (kinetic energy). In order to study a given problem of dynamics we can consider the quantity of motion as determined by force (Newton's laws) or, alternatively, find the vis viva through the work done with the body. Mathematically it is extremely simple to show that Newton's second law, with a small algebraic manipulation, can be transformed into the principle of work and of kinetic energy.

These two forms of approaching a problem of dynamics were separated for a long period and caused a celebrated controversy, which we have already mentioned. France and England adopted the Newtonian path until the problems with dynamics raised by machines changed the course and the trajectory of these studies. A large part of this effort was due to Lazare Carnot. The development of this path by polytechnic engineers would be shown to be fruitful to the extent that it organized a new discipline of engineering, applied mechanics, but it would prove incapable of taking into account the entire complexity that a machine in movement represented with a new type of exchange, now between work and energy in the form of heat.

A final observation about Part II. Being more concerned with a history of applied mechanics where the principal characters are renowned figures who frequently appear in engineering manuals, it represents a modest contribution to the history of mechanical engineering. It should be highlighted that science history texts dealing with the more applied sciences are very rare. Normally they are histories of the basic disciplines, such as mathematics, physics, chemistry or biology.

To finalize these comments, let us consider a type of counterpoint made in some parts of the text on the positivist philosophy of Augusto Comte. This has to be done due to our profound disagreement with this current of thought, which together with neopositivism constitute an important current of the philosophy of science. Our points of divergence are many and are explained throughout the text. These can be summarized in the positioning of positivism in relation to reality, in other words, in its relativism, in its attempt to reduce the social sciences to a social physics, and finally in its continuous 'evolutionary' perspective of knowledge and the development of science.

Obviously, it would be impossible to deal with these questions in the depth that they deserve in this concluding part of our investigation. However, to omit it would be worse.

For Augusto Comte, science is not knowledge of the whole world and the collection we have of scientific, empirical, knowledge etc., is only an approximate form of ordering the world. Human knowledge for him is limited and relative, and to a great extent depends on an evolution that has yet to occur. Comte thus adopts a relativist position. According to him no philosopher, no logic or general axiom can guarantee a system of knowledge of a unified reality. Only indirectly, a posteriori, through the internal coherence of different methods is reality revealed in the history of particular sciences (Grange 1996).

Marxists also consider knowledge to be a form of approximating reality with respect to its historical nature. The role of science is to construct the theoretical, experimental and empirical instruments in the best way for it to be accelerated, though without any metaphysical skepticism.

In relation to the project of transforming sociology into social physics, the difference between Marxism and positivism is abyssal. Given the difficulty of carrying out any historical analysis without considering opinions and value judgments, positivism opts for a type of squaring the circle, in other words, it looks for social and historic facts in a pure, aseptic state, without ideology and stripped of value judgments. Is this possible? Is it not a precursory idea of the structuralisms of the twentieth century?

In relation to the evolutionary character of knowledge, in a different meaning from what we use in this study, and one in which successive stages of knowledge are always reached, as the model of scientific progress, we should say that this idea is at the core of concepts that affirm we are living in a society of knowledge heading towards unimaginable and irreversible conquests. We will turn once again to Enrique Leff (2000). Although in a different context, criticizing the current civilization as being one of knowledge, he states: "However this civilization of knowledge is, at the same time, a society lacking knowledge, of generalized alienation, of the uprooting of wisdom and the disenchantment of the world (the dead poets' society; a society without purpose, without imagination, without utopia, without a future)".

To what extent can we say that we live in a knowledge society? Also to what extent is scientific knowledge cumulative? What we are certain of is that never in history were there so many humans who knew so little and who were so isolated, marginalized and expropriated from decision making processes related to their conditions of existence; there was never so much subjugated wisdom, so much devalued and under used knowledge due to the precariousness of labour and living conditions; never were traditional knowledge and wisdom so uprooted from their natural environments and their territories, so expropriated from their cultures and identities.

The history of science has shown the exact opposite and our investigation reinforces the point of view of a development of a discontinued and apparently erratic science, where the historian has to investigate the general movements that propel scientific knowledge, the secondary flows and reflows that mix with the principal caudal, identify the general picture, the private interests and motivations to understand reality in its complexity and through this to construct a coherent and rational discourse of its evolution. It is not by chance that the three most eminent epistemologists and historians of science in the twentieth century, Alexandre Koyré, Thomas Kuhn and Gaston Bachelard had a discontinuous view of scientific development. Koyré's has already been mentioned above. In relation to Kuhn, we can say that his model of progress which affirms that a scientific theory is replaced by another through a scientific revolution in which normal science gives place to a new scientific paradigm contains nothing continuous nor gradual, much to the contrary. Bachelard (1996) proposes the construction of an approximate knowledge which is also discontinuous: "Even in the historic evolution of a specific problem there appear real ruptures, brusque mutations, which overcome the thesis of epistemological continuity".

We will now move on to three questions which I promised to deal with in a succinct form at the beginning of this chapter. The first is related to the cultural changes that have occurred since the fifteenth century, or a little beforehand in Europe, in the sense of the acceptance of practical arts and of manual labour as an important stage in the seventeenth century Scientific Revolution. Paolo Rossi's book, *Os Filósofos e as Máquinas*, authoritatively analyzes the theme and partially fills this gap.[1] The second question is the problem of technique, although there

[1] Rossi, P., "Os Filósofos e as Máquinas", Editora Schwarcz Ltda., 1989, S. Paulo (1989).

exist numerous renowned texts about this subject. The suggestion for the deepening of this theme was first given by Lucien Fèvre in an article in the journal he founded with Marc Bloch in 1929, the *Annales d'histoire économique et sociale*, no. 36, 30 November 1936, entitled: "Reflections on the History of Techniques".[2]

Fèvre suggests that a history of techniques be looked at from three distinct points of view: (1) what he calls a 'technical history of techniques' which covers the technical study of procedures, instruments, and various technical activities linked to their epochs; (2) a study of these sets of procedures, tools, manufacturing activities, and how they are affected by a history he calls 'evolutionary,' and how progress causes problems related to their achievement, which he describes as problems of relations between 'theory and practice,' between 'science and technical invention,' and between 'chance and necessity;' (3) a study of the relations between technical activity and other human activities (such as religion, art, politics, etc.).

The third question raised in our text is the study of the machine as a social artifact, following the suggestion of Harry Braverman, which was quoted in the epigraph in the Introduction. However, it is not only for this reason, but mainly because the 'context of machines' requires its deepening. This discussion will try to answer questions like the following: What are the fundamental causes of the expansion of machinism? How and why did the machine replace human labour? What are the impacts of the expansion of machinism on labor relations? Although our concerns are not restricted to this brief questionnaire, we will try to complement it in our text with some reflections. Due to its importance for the investigation, we will pay special attention to the third topic.

Between the fifteenth and sixteenth centuries Europe underwent a radical transformation of how the mechanical arts and manual labour were perceived. This can be seen in the work of Leonardo da Vinci[3] and later of Benedetti (1530–1590),[4] Galileo and others, where mechanics came to be seen as the most noble of sciences. Some historians have called this change a 'return to Archimedes.' In the wake of this process a cultural revolution was underway with various impacts on the social and economic components of the period.

European culture was brought by these events to a definitive rupture with the thesis of the inferiority of technique in relation to science and consequently of a similar relationship between manual work and intellectual work, a characteristic which had been part of classical civilization, both Greek and Roman. The origin of this opposition was in the economic structure of a slaveholding type society which counted on an 'abundance of living machines' as well as dispensing with and

---

[2] Cited by Simondon G in "L'invention dans les Techniques", p. 32, Editions du Seuil, 2005, Paris (2005).

[3] See Carlo Pedretti, *Leonardo: The Machines*, Editora Giunti, 1999, Florença (1999).

[4] In some ways Benedetti is a continuator of Leonardo da Vinci, as well as being a critic of the work of Aristotle. His work was published in 1585 as *Diversarum Speculationum Mathematicarum et Physicarum*, and covered practically all the branches of mechanics at that time. See René Dugas, "History of Mechanics", p. 104, Dover Publications, Inc., 1988, New York.

making superfluous the construction of machines able to replace human labour. From this arose the contempt in which slaves or anyone who carried out manual labour was held, as is clear in the manner in which the two greatest thinkers of antiquity, Plato and Aristotle, treated manual labour.[5]

This centuries old antithesis would collapse during the Renaissance where there emerged various formulations aimed at uniting manual labour with the spirit of scientific activity then in frank ascension. Over time this movement acquired different forms in the various European countries. We are referring to the distinct nuances which the Enlightenment would assume in the countries of Europe.

When Hugo da Gama Cerqueira in his interesting article, "Adam Smith e seu Contexto: o Iluminismo Escocês,"[6] deals with the case of Scotland, however, he looks at the new way in which the most recent historiography has seen the phenomenon of the Enlightenment: "Rejecting the monolithic and anachronistic patterns of this type of approach, the most recent historiography on the Enlightenment has been concerned with the diversity of its national manifestations and the particularities of these different expressions which, in the conventional readings, were obscured or simply taken as signs of their insufficiency in the absence of the 'true' Enlightenment. In fact the presence of the Enlightenment in England was fully felt in the seventeenth century: the flourishing of reason, respect for freedom of expression and religious tolerance were recognized and celebrated by the foreigners who visited the country." In the same text Cerqueira refers to the particularities of the Enlightenment in England, stating: "The constant search for improvements, the application of science and the mechanical arts to practical purposes, the belief in trade as the promoter of tolerance and social cohesion. All of this shaped a British path to the Enlightenment, distinct from the variants followed on the continent by its markedly individualistic accent and by its "effort never to subvert the system, but to protect it, in order to achieve individual satisfaction and collective stability within the post-1688 structure."

In his article, "A Idéia de Ciência Aplicada na Inglaterra do Século XVIII: os Manuais Newtonianos,"[7] Luiz Carlos Soares, also looks at the English Enlightenment from an original and little researched point of view, namely the diffusion of scientific knowledge based on the mechanics of Newton[8] through itinerant teachers. According to Soares: "The idea of Applied Science, related to the needs of industrial activities and the welfare of the population of the country,

---

[5] See Joel Jung's book: "Le Travail", Editora Flamarion, p. 111 and p. 143, 2000, Paris (2000).

[6] See Anais do Seminário de História da Ciência e da Tecnologia, October 2005, B. Horizonte (2005).

[7] See Anais do Seminário de História da Ciência e da Tecnologia, October 2005, B. Horizonte.

[8] Book II of Newton's *Principia* deals with the movement of bodies in a resistant environment, providing a theoretical foundation and suggesting broad possibilities for the application of Newtonian mechanics. It is divided into nine items, with the first three studying the movement of bodies in a resistant environment in which the forces that are opposed to these movements depend on a function of velocity. Item IV studies circular movement in resistant environments, item V studies hydrostatics, item VI studies the movement of pendulums with resistance, item VII studies the movement of fluids, item VIII studies the propagation of waves in fluids and finally item IX studies the circular movement of fluids.

constituted one of the most important aspects of the English Enlightenment and a powerful intellectual lever which allowed the emergence of the Industrial Revolution from the 1780s onward."

Returning to the cultural transformations in Europe, a synthesis of the two opposing movements can be found in Galileo. Only in his work did the full convergence between the experience and practice of artisans, technicians and builders of machines and the theoretical tradition of European science occur.[9] It is important to emphasize this, since some of those who have studied Galileo's work[10] have divulged a distorted image of him, presenting him as a rationalist and pure mathematician and what is worse someone even opposed to the knowledge of Renaissance engineers and practitioners. It is also true that Galileo was fully aware of the role and the importance of preparing a scientific theory that would move beyond technical and empirical reports, observations and records, giving them a generalizing nature through abstraction.[11]

Another important figure in this cultural transformation process is Francis Bacon (1561–1626), a representative of the movement that replaced a literary-rhetorical culture with a technical and scientific one.[12] He also recognizes that to carry out these changes social reform is necessary which implies a rupture with the tradition then in force, and that this involves not only changing the mode of thought, but also people's way of life. Above all, a new attitude towards the natural world has to emerge.

---

[9] In his *Discorsi,* in the first dialogue of the first journey, Galileo, expressing himself through Salviati, who represents him, states: "It appears to me that the frequent activity of your famous arsenal, my dear Venetians, offers a vast philosophical field to speculative intelligence, particularly in that subject called Mechanics, since in this place all types of instruments and machines are continually built by numerous artisans, amongst whom it is possible that there exist, due both to observations made by ancestors, as well as those who continually make their own reflections, some who join reasoning" (Galilei 1988).

[10] See Alexandre Koyré in "Galileu e Platão", Editora Gradiva. On page 12, he states:their expertise to a profound "The science of Descartes—*a fortiori*, that of Galileo—is nothing other (as has already been mentioned) than the science of the artisan or the engineer. I must confess that this explanation does not seem to me to be entirely satisfactory. It is true, as I understand it, that modern philosophy, such as modern religion and ethics, put the tonic in action, in praxis, much more than ancient and medieval thought. The same is true in relation to modern science: I am thinking of Cartesian physics and its comparisons with pulleys, ropes and levers. However, the attitude we have just described is much more that of Bacon—whose role in the history of science is not of the same order—than of Galileo and Descartes. Their science is not that of engineers or artisans, but of men whose work rarely left the sphere of theory" (1963).

[11] In his book *Galileu Galilei,* Editora Nova Fronteira, 1997, p. 335, Ludovico Geymonat states: "No matter how important a function Galileo attributes to mathematics among the phases of scientific research, examined in the two preceding paragraphs, what he attributes to it in the construction of physics theories is undoubtedly superior: it is so relevant that it created the conviction in some interpretations that Galileo intended to reduce scientific theories only to their mathematical aspect" in Chap. 4.

[12] See the book by Bernardo Jefferson de Oliveira: "Francis Bacon e a Fundamentação da Ciência como Tecnologia", Editora da UFMG, 2002, B. Horizonte (de Oliveira 2002).

Bacon argued that the philosophical discourse of the classical world based on the superiority of contemplation about what to do and of resignation in relation to nature had to be overcome. Instead he proposed the conquest of nature. This was the rupture he wished to announce.

In his previously mentioned book, Paolo Rossi cites some of the eminent figures of the Renaissance whose work contributed to this change of attitude in the cultural scenario: "Palissy, Norman, Vives and Rabelais—at different levels and with different intentions—had expressed the demand that was widely diffused in the sixteenth century for a type of knowledge in which the observation of phenomenon, attention to works, and empirical research would be more important than rhetorical digressions, verbal complacencies, logical subtleties, and a priori constructions"[13]

Bernard Palissy, a French ceramist, stated that he learned more natural philosophy from a few hours of practical work than reading from philosophy books for hundreds of hours. Robert Norman, an English seaman, describes his knowledge of the magnetic needle and its movements and confesses that he considers himself to be entirely incapable of holding a discussion with logicians. However, his publication and experiments would benefit England. Luis Vives, a Spanish humanist, valorizes techniques and appeals to intellectuals to study agricultural, navigation and machine construction techniques. Only in this way would they know how to use these for the benefit of all, since what philosophers had constructed were metaphysical and imaginary entities. He called on educated men to abandon their disdain for vulgar knowledge and to start to frequent the workshops and factories to learn with artisans the practical arts. Finally, the renowned French writer François Rabelais (1494–1553) considers the knowledge of artisan practice as fundamental for the education of youths. His character Gargantua had to concern himself with traditional studies, but also saw how metal workers, the goldsmiths, cutters, watchmakers and printers, etc., worked.

In summary, without the ascension of the cultural 'status' of experimental philosophy, it would have been difficult for science to have developed, since this would have required the help of these two forms of knowledge, as well as the crucial and decisive task of building instruments to test hypotheses postulated by scientific theories.[14]

Let us now look in a general manner at some important questions related to technique, the second question raised at the beginning of the chapter. In 1955 Jacques Ellul's (1912–1994) book entitled *La Technique ou l'Enjeu du Siècle* was published in France. It was only published in Brazil in 1968, translated and prefaced by the much missed Roland Corbisier (1914–2005). Over time this book has become one of the best known and cited references in studies of techniques. Ellul combines an original analysis with disconcerting comments and often puts well-known authors in disagreeable situations. In his long preface, Corbisier calls the reader`s attention

[13] See the previously mentioned book by Paolo Rossi, p. 25 (1989).

[14] See Pierre Thuillier's interesting article: "Au Commencement Etait la Machine", La Recherche, Janeiro de 1976, p. 47 (1976).

to the fact that Ellul had undergone a considerable turnaround in his thinking and had deeply shaken many of his long crystallized convictions about techniques.

One of these certainties which Ellul's book questions is the identification which many famous authors make between technique and machine.[15] His analysis is based on his affirmation that although machines are "technique in a pure state," since technique tends to mechanize or machinize everything, "it is the machine that currently utterly depends on technique." According to him, technique is something much more wide-ranging and is a demand of rationalization, a superior form of 'knowing how to do,' and assumes important characteristics in sectors and human activities where the machine does not function.

In his analysis of Ellul's book, Corbisier highlights another question which has generated much discussion. This is the somewhat widely disseminated statement that techniques involved an application of science, which Ellul considered false. He argued that technique preceded science and primitive magic, which has nothing to do with science but which is still a technique. Only very recently has this relationship been inverted and although science contributes in a decisive form to technological progress, it is totally subordinated and dependent on technical progress, becoming what Ellul calls an "instrument of technique."

Ellul attributes the following characteristics to techniques: automatism, self-growth, indivisibility or individuality, universalism and autonomy. Let us look at what each of these concepts means. By automatism he understands the impossibility of refusing a solution or method involving a higher level of technique. In other words, the technique imposed is always the one that is more rational and more effective. According to this characteristic of technique, man is no longer the subject of the choice between these alternative techniques, rather it is the technique which makes the choice and thus it acquires its own automatism.

In relation to self-growth, Ellul notes that technique tends to develop, to progress, increasingly excluding human intervention, while technical solutions engender each other, in a chain process which becomes more and more automatic. It is because of this, he argues, that the same discoveries occur simultaneously in different countries with equivalent levels of technological development. Thus, the incentive driving technical progress is technique itself, creating a system with the capacity to reproduce itself.

He calls the third characteristic indivisibility, individuality, or also uniqueness. No matter what name is chosen, it signifies that the technical phenomenon by encompassing the entire set of techniques *ipso facto* constitutes a totality which is always presented with the same characteristics. As a result it is difficult to distinguish between the technique and the use to which it is put, since the essence of the technique consists of its use, which in principle is neither good nor bad, just or unjust, because as it is a technique this is the only possible use, and it cannot be judged by non-technical criteria.

---

[15] Ellul principally criticizes Lewis Mumford. See *La Technique ou l'Enjeu du Siècle*, p. 2 (1968).

This line of argument developed by Ellul leads to another question which is that it makes no sense to try to guide techniques by ethical standards or subordinate them to any type of logic, which in other words, leads to the legitimation of any use that is made of them. In short, technique and the use which is made of it coincide. Otherwise it involves claiming that technique is no longer technique.

The characteristic of universalism is something much more wide-ranging, is identified with the actual process of the universalization of western technology, and can be looked at from two different perspectives: the geographic and the qualitative. The latter is a transformation which operates on the former. Generally speaking technique penetrated and spread throughout all countries, in all the regions of the globe. All those who neither developed it nor assimilated it to the necessary extent and scale wanted and demanded it. This process led all peoples and countries to follow the same path, although there were enormous differences in the stages followed by each of them. However, each followed the same trajectory.

Finally, we will examine the last characteristic of technique, which is autonomy. This signifies that technique tends to disconnect itself from all traditional authorities, powers, or forces and has its own laws of development, respecting nothing which is foreign or hostile to it. This autonomy is something Promethean in its thirst and demand for infinite power and domination, becoming an absolute force which is enough in itself and does not depend on any ethical judgment, being situated above good and evil. It does not tolerate conflict with any other value, much less be judged by anything else. It thus becomes a supreme value. All other values should be judged as a function of it and not to the contrary.

We have tried to reproduce as faithfully as possible the central core of Ellul's ideas contained in the above mentioned book. Although the role and weight he attributes to technique in society is overwhelming, we do not know which internal mechanisms and which contradictions within objects or technical phenomenon move this system. However, Ellul's work on technique should be studied and more especially responded to. In reality it consists of a trilogy. In addition to the book in question there are *Le Système Technicien*, published in 1977 and *Le Bluff Technologique*, from 1988.

In 1977 in his second book on technique, Ellul expanded and deepened his previous analysis made 20 years before, now seeing technique as a system. He agrees that nowadays technique has a configuration which can be perfectly classified as a system in its traditionally understood meaning, although it has a set of specificities. In defining a system he states: "A system is a set of elements, some in relation to others in such a way that the evolution of one of them provokes the evolution of the set and that modification of the set has repercussions for each element."[16] This signifies for him a network of inter-relations. In addition to these considerations, Ellul sees an additional complexity in the composition of constitutive elements of the system. Some of these can have a quantitative nature while others do not. As a result the speed of changes of each of the factors is not identical. In other words,

[16] Jacques Ellul, "Le Sistème Technicien", p. 88, Editora le Cherche Midi, 2004, Paris (2004).

the system has its process and its speed in relation to the specific changes that occur in its parts. Another characteristic which Ellul considers fundamental is that in the interior of the technique system, its elements have a certain preferential aptitude and combine among themselves before entering in combination with the external factors. Moreover, this system behaves in the inter-relations between the parts differently from many other systems, such as mechanical systems, whose dynamics are repetitive and entirely predictable. In the technique system the modifications that its elements undergo modify the other elements and its action is never repetitive, but constantly innovative. The system is in permanent evolution and never returns to a previous state.[17]

The third question we will look at is closer to our study. This involves understanding the process of expansion of machinism, or, as we also suggest at the beginning of this chapter, trying to answer these questions, which basically means determining the contradictions and the dynamics of this expansion in the various contexts in which the physical concept of work was developed. As has also been seen the expansion of machinism is a fundamental fact of modern capitalist industrial society. In the rest of this chapter we will use the ideas developed by Marx related to this question, which principally appeared in the first volume of his fundamental work *Das Kapital*. However, before this we will make a brief detour.

In 2005 Daniel Romero published the book *Marx e a Técnica*[18] arising out of a master's thesis in UNICAMP, whose principal merit was to bring back for discussion Marx's ideas about the relations between science, technology, capital and the relations of labour. He also carries out a meticulous inventory of the works of Marx concerned with this question, the evolution of his ideas and the concepts created by Marx to study the problem. An interesting contribution of Romero's book is the 'discovery' and the use of Marx's 1861–1863 manuscripts.

Romero highlights in his book two readings made by Marx which influenced him greatly: "Only in 1845, when he was in Brussels, did Marx look at the question in a form which would be fundamental for his later conceptions: he began the studies of *On the Economy of Machinery and Manufactures* by Charles Babbage,[19] a professor from Cambridge University and *Philosophy of Manufactures*, by Andrew Ure,[20] an English chemist." Both books are quite representative of the period of the expansion of machinism in England. Babbage's book

[17] Ellul calls attention to the fact that a technique system contains a large margin of probability and that there is no point in making predictions of 'inventions' in technique, since predictions are only possible as functions of a global study, while it is also a system much more difficult to study than the physical, ecological, etc. He also notes that it is not useful to carry out computer simulations of such a system without first constructing a complete theory of it.

[18] Daniel Romero, "Marx e a Técnica", with the subtitle *Um estudo dos Manuscritos de 1861–1863*, Editora Expressão Popular, 2005, S. Paulo (2005).

[19] Charles Babbage, *On the Economy of Machinery and Manufactures*. In this work I have used the French translation: "Economie de Machines et des Manufactures", Librairie du Dictionaire des Arts et Manufactures, 1880, Paris (1880).

[20] We also used a French translation of Andrew Ure`s book: *Philosophie des Manufactures et Economie Industrielle*, 1836, Bruxelas (1836).

acquired enormous fame, not just in England, and is directly related to our theme, for which reason we will make a small incursion into it. We will also have the opportunity to verify which discussions guided Marx's studies, in order to compare them with what was being developed by the polytechnic engineers in France at the same period. Although Ure's book contains a long chapter dedicated to the questions of manufacturing in general, it is mostly concerned with the British textile industry, though it contains important information about French industry.

In 1831 Charles Babbage, an English mathematician who invented a calculator, published the above mentioned book, presenting an original analysis of industrial development in England, as well as containing statistics and comparisons between English and French industry. The book has 460 pages and is written in simple and direct language which is easy and pleasant to read, with short and objective chapters. It covers a very wide and varied range of questions linked to manufacturing, including the organization of workers. It is divided into two parts: Economy of Machines and Economy of Manufacture.

On the first page of the first part, Babbage summarizes the objective of the book, stating that it involved "the mode of action of tools and machines and, includes a set of mechanical considerations to evaluate them." In the second part he "presents a profound discussion of various questions of political economy which are linked to the manufacturing system. We will next look at the conditions of success in the use of machines and subjects closely linked to the interior organization and existence of large industrial establishments; finally we will study their relations with the general interests of society." He closes the book analyzing the influence of science on the future development of industry.

Obviously, it is not our objective to produce an in-depth study of Babbage's book, as we did with Lazare Carnot's *Principles* or also with the works of Navier, Coriolis and Poncelet. We only wish to complement our vision of questions related to the physical concept of work discussed during this investigation and to extend them to England.

An important fact mentioned by Babbage is that after the fall of Napoleon (1815), communications between France and England were restored after a long period of war and it is only after this that is it possible to make comparisons between their industrial capacities. He adds that it was precisely in the 1820–1840 period that the generalization of industrial capacity occurred in France and on the continent. He also warns about the nascent America which has "immense resources and an entire continent" and whose position was leaving its rivals in the Old Europe nervous. It should be taken into account that it was in the period between the end of the eighteenth century and the first three decades of the nineteenth that the incorporation of the physical concept of work in the economy occurred as seen in the previous chapter. Babbage's book thus assumes enormous importance and needs to be analyzed if we want to deepen and enrich the conclusions of our investigation.

Babbage specifically deals with the physical concept of work in a machine on page 17 where he states: "If we analyze the operations carried out by any industrial manufacturer, we can clearly see that this always involves a worker, who has to make a certain effort along a determined path, due to the nature of this operation. These two elements, effort of a certain length and the path followed of a determined nature, are precisely what machines aim to effect more economically than the direct labour of man, and in such a way that reduces as a consequence to a much greater proportion of manufacturing costs." Without defining or mentioning the product of the force by its motion from its point of application as 'work' he emphasizes that it is this quantity which matters in operations carried out by a machine independent of human labour in the operation of the machine. It is also the same quantity which should be optimized from the point of view of the reduction of manufacturing costs. Babbage does not develop any theory of how this optimization could be done, but demonstrated that the concept of work is the axis around which the fundamental questions of manufacturing turn. For him the concept of work was already incorporated in economics.

Two observations made by Babbage demonstrate that he was aware of what was happening in France and was monitoring the publications of the French polytechnic engineers. On page 18, referring to the work of a machine, he states: "it has been proved by experience that in practice everything that is gained in time is lost in force," a proposition that was very well known at the time, and already stated by Lazare Carnot, as we have seen previously. On the same item, referring to the great progress made in the teaching of mechanics, he cites the efforts of Poncelet in this area. As we have seen, Poncelet had been assigned by the French government the task of creating an applied mechanics course in his native city, Metz.

On page 19 at the end of Chap. 2 he states: "When we raise a 1 kg weight to a height of one meter, the effort is evidentially 1 kg, as well as when we raise it to two meters; but the effort used, the labour, that which, as we have said, is paid, is evidentially the double in the second case: it is the product of the effort by the path followed, for which the kilogram unit was created, all forces can be calculated in kilograms and all lengths in meters.

It is this quantity that remains constant, except for the losses due to friction, of one ball bearing against another in the machine. The path run and the force can vary at any instant, but its product, work, always remains constant. It is these variations of efforts and paths which the arm of man guided by his intelligence carries out almost without perceiving (after a long apprenticeship), that the aim of machines is to achieve the almost gratuitous transformation of raw materials into useful objects for our needs and our pleasure." Some observations need to be made here. First, at no moment does Babbage mention the expression vis viva to designate the product of the force by the velocity squared, despite the fact that it refers to the total capacity of the machine to carry out work. This means that work is already separated from vis viva and identified as a determined quantity, which was not true in the mechanics of Lazare Carnot. Babbage also used the scheme that was common at the time of measuring the capacity of a machine to carry out work by obtaining the product of the weight it is capable of lifting by the height to which this weight is lifted. Finally, he refers to useful work, or net work. In other words, this work divided by the total work given to the machine will provide its productivity. In summary, the concepts, the comparative scheme, the concerns with effectiveness and the reduction of manufacturing costs, are all similar both in England and in France.

In the second part of Babbage's book many questions which are at the center of our discussions are raised. In Chapter eighteenth, which has the suggestive title "The price in silver considered as a measure of the value of things," we can read the following: "The price in silver paid for any object cannot offer us anything other than an inexact comparison of various values of this object, at various times and in different countries; since the gold and silver which serve to measure this price are subject to variations like any type of merchandise and do not offer in any way a constant base which can be used for similar comparisons. In this invariable base we believed we had found the average price of certain types of raw or manufactured products." Next, continuing the discussion of the value to be attributed to a given merchandise, he lists a series of difficulties with taking the average price of some raw products as an invariable quantity which translates the value that is desired. Babbage adds: "After my own observations, that of all the causes which can influence the reduction of prices, the most energetic, without a doubt is the invention of more economic manufacturing procedures".[21]

The problem for Babbage was the same as the one polytechnic engineers faced in France. This involved finding a form, either material—as in a type of merchandise—or immaterial such as the concept of work, which could act as an equivalence between different merchandise and consequently allow the establishment of a process of exchange in the market. The polytechnic engineers, as we have seen, tried to use the concepts of mechanics. Babbage would continue to apply them throughout the economy. It should be highlighted that in this stage of development it did not make much sense to make a distinction between the physical and the economic concept of labour. When the physical concept was applied to the economy it acquired and incorporated economic characteristics. We can call it the physico-economic concept of work. The central question continues to be how to find equivalence. For us, however, understanding how this happened and that the concept of work which emerged from mechanics was at the center of the debates is fundamental.

Babbage also perceived the relevance of the role of technology as the most important factor in the reduction of prices. He continued to search for a form that would allow the comparison of goods and states: "From these discussions it appears that there does not exist any material, not even a combination made with various industrial products, which can provide an invariable unit to establish a scale of comparisons between the values of the same good in different epochs. Malthus proposes using the work done in one day by a worker for the unit to which all other values will refer." He then continues: "For similar comparisons there also exists an element which is not rigorously necessary, but which sheds an

[21] Babbage, op. cit. p. 188.

important light on the question; it is an exact appreciation of the quantity of nutritive substances usually consumed by the worker, comparing the indispensible quantity for his daily sustenance with what he is paid allows him buy."[22] He continued his discussion of the problem of value, resorting to Malthus to bring into the debate the working time in a daily work day. Furthermore, although he explores an energetic approach, he addresses the central question which is the remuneration of the reproduction time of the labour force. Marx would unveil the mystery, talking about a surplus time for which the worker is not remunerated.

Another relevant question which Babbage also discusses is the division of labour: "Of all the principles of the manufacturing economy, the most important perhaps is the division of labour between the individuals who compete to prepare the manufactured product".[23] He continues: "The first causes of the general advantages which result from the division of labour have been the object of great discussion among the authors who deal with political economy; however, it does not seem that the relative importance of the influence of these various causes has been appreciated in all cases, with the necessary precision. I will thus quickly discuss these first causes."[24] We will not enter into the merit of this discussion, but only register the importance which Babbage attributed to it.

Finalizing this brief overview of Babbage's book, we have to highlight that like the classical economists he also believed that the replacement of workers by machines would not increase unemployment, since by lowering the price of goods there would always be a need for new contingents of them to meet this increase in demand.[25]

This was the situation and the agenda of the debates under way when Marx came onto the stage. He had written the *Economic and Philosophical Manuscripts*, published in 1844, and read in addition to the books of Babbage and Ure, as mentioned above, the work of Engels, principally *The Situation of the Working Classes in England*, released in 1845 and had worked his way through the economic writings of Jean Baptiste Say and Adam Smith.

Marx analyzed the question of science and technology within the framework of the contradictions which arose out of the capital-labour relationship, which signified studying the exploitation of workers by capitalists, as well as a specific and improved method for the extraction of relative surplus. The expansion of machinism is for him one of the forms of the devaluation of labour to the extent that it reduces the prices of goods in general, including that of the labour force, as well as managing to impose great control and rhythm over production. In summary, the

---

[22] Babbage op. cit. p. 190.

[23] Babbage, op. cit., p. 198.

[24] Babbage, op. cit., p. 199.

[25] See Cláudio Napoleoni: "Smith, Ricardo, Marx", Edições Graal, 6ª. Edição, 1988, Rio de Janeiro (1988).

machine imposed a type of capitalist rationalization aimed at exploitation and the dominion of production. In the phases of the accelerated industrialization of capitalism the type of rationalization imposed by the entrance of the machine into production gained a specific form of the valorization of capital.

These preliminary observations highlight that Marx refuses a general or universal history of technology aimed at understanding its development beyond the social formations of each epoch. If this general history existed we would determine to a great extent historical development itself. An inversion and a mistake that in our opinion is committed by Jacques Ellul, which compromises the brilliant analyses he makes of technique. Other authors also share this point of view and intend to find in technological progress the cause of the general development of society. These perspectives of scientific and technological advances start with the hypothesis that the productive forces are neutral in relation to the relations of production. From this there arises another conception according to which development in general and progress in the scientific and technological field consist of a continuous improvement of the techniques of production independent of economic formations. Marx`s analysis, by subordinating scientific and technological development to the specific forms of economic production, signifies that the overcoming of the current mode of capitalist production implies another material foundation for society and consequently another form of technological development.

Marx's analysis of machinery is found in the first volume of *Das Kapital*, in Chapter 13, p. 425. He starts with a citation from John Stuart Mill: "It is doubtful that the mechanical inventions discovered until the present have relieved the daily toil of any human being" and notes that Mill had refrred to "some human being who does not live from the work of others," adding "machines certainly have increased the number of idle wealthy." Marx complements his reasoning by stating that when capital expands machinism, it does this with the aim of making goods cheaper and in this way reducing the part of the working day dedicated to the labourer himself. By proceeding in this manner he is expanding the other part of this same working day which he gives free to capitalism. The introduction and expansion of machinism is the most efficient means to produce and extract surplus from workers.

From what has been seen above, we can draw the conclusion that the method used by Marx to analyze the expansion of machinism in industrial production assumes the existence of specific forms and relations for the different social formations, with it being assumed that it is even natural that it has these specifications during the various phases of industrial development. In *Das Kapital* he analyzes the details, the transition from manufacturing to modern industry and the nuances of this process, underlining that in manufacturing the starting point to revolutionize production is the labour force, while in modern industry it is the instrumental labour which is transformed from a manual tool into a machine, thereby establishing the difference between machine and tool. Beforehand, however, he defines what a machine is.

Similar to the French polytechnic engineers, Babbage, and many others, Marx considered a machine to consist of three parts: the engine, the transmission, and the machine-tool or working machine. Obviously, what he calls a machine-tool should not be confused with the modern concept machine-tool, but simply is the part of the machine directly responsible for the mechanical work and which achieves what the machine was designed for.

After the development of the machine Marx makes an original analysis of the various phases of the Industrial Revolution. According to him, it is tools that started the Revolution. Moreover, tools are also responsible for the transformation of a workshop or manufacturer into mechanized exploitation. Up to a certain point this is surprising, since the majority of historians identify the engine of the initial transformation as the replacement of human or animal efforts by a natural source to supply energy, notably the steam engine.[26] According to Marx: "When the tool in a strict sense is transferred from man to a mechanism, the machine takes the place of the simple tool. The difference is really striking even when man continues to be the primary engine. The number of tools which a man could operate at the same time is limited by the number of his natural instruments of production, his physical organs." And he continues: "The number of tools which the machine-tool operates simultaneously is emancipated, from the beginning of the organic barrier which the manual tools of a worker cannot surpass".[27]

In relation to the role of the steam engine, Marx wrote: "The actual steam engine in the form it was invented at the end of the seventeenth century, during the manufacturing period, and which lasted until the 1780s, did not cause any industrial revolution. To the contrary, it was the creation of machine-tools which made a steam engine revolution necessary. When man comes to use only the driving force of a machine-tool, instead of using the tool on the object of labour, wind, water, steam, etc., can take his place and the use of human muscular strength as a driving force becomes accidental".[28] Continuing his analysis Marx states: "The machine from which the industrial revolution arises replaces the worker who handles a single tool with a mechanism that at the same time operates with a certain number of identical tools or similar to it, and is triggered by a single driving force, no matter its form. We have the machine, but still as a simple element of mechanized production".[29]

---

[26] In his *Revolução Industrial*, Publicações Europa-América, 2ª. Edição, pp. 93–94, T. S. Ashton refers to the steam engine as follows: "The new form of energy and transmission engines, which powered engines which previously had been moved by muscular strength, were the path through which industry entered the modern age." Another renowned author and historian is Paul Mantoux, who wrote *A Revolução Industrial no Século XVIII*, Editora da Unesp/Editora Hucitec. On p. 333 he states: "Until then the steam engine was nothing other than an improved fire engine. It was used for this purpose in mines or in water services. With the invention of circular movement, it became a driven machine: its uses hereafter could indefinitely expand; the entire industrial field opened for it." On p. 339: "This capital fact, the advent of the steam engine, opened the last and most decisive phase of the industrial revolution. Freeing large-scale industry from the barriers which still weighed on it, steam allowed its immense and rapid development" (Ashton 1995; Mantoux 1988).

[27] Das Kapital, p. 427.

[28] Das Kapital, p. 428.

[29] Das Kapital, p. 429.

In Marx`s analysis the steam engine is revolutionized by the constant changes which occur in the tool system of machines, improperly called machine-tool, and which, as we have seen, is responsible for the operational part in the strict sense of the machine. This requires a more powerful trigger (engine), one which can surpass human capacity. Added to this is the irregularity of manual switching in terms of the need for uniform and continuous motion. The evolution of tool systems leads to its most developed system when it comes to compose an organic whole, which Marx describes as follows: "Mechanized production finds its most developed form in the organic system of combined machine-tools which receive all their movements from a central automaton and which is transmitted to them by a transmission mechanism. There, thus, emerges in the place of the isolated machine, a mechanical monster which fills entire buildings and whose demonical force is disguised in the almost solemn rhythmic movements of its gigantic members and erupts in the feverish flurry of its innumerable work organs".[30]

In order for an expansion of machinism to occur, the machine itself and its form of use had to be revolutionized. Previously it had consisted of an isolated unit moved by a single man, what emerged was a system of machines, whose origin was in the revolution that occurred in tools, though this stage could not have been reached before the steam engine replaced the previous driving forces. This is the real revolution caused by the steam engine, in other words it allowed for a system of machines to appear on stage. Marx described this process as follows: "Moreover, at a certain stage of development modern industry technically entered into conflict with the base it possessed in artisanship and manufacturing. The widespread growth of the sizes of engines, the mechanisms of transmission and machine-tools".[31]

This is the internal dynamic of the process of expansion of machinism, in other words, when it is analyzed from the point of view of the factory or a unit of production. When we observe this process and its diffusion through the various types of industry, it can be seen that it induced a revolution in the systems of transport, communication, and others, resulting over time in significant changes not only in production as a whole, but in all of society.

To finalize this brief analysis of the third theme proposed at the beginning of this chapter, we will return to the question of value, drawing for the last time on Marx and his description of how much value a machine can transfer to the product. He states the following: "From the exclusive point of view of making the product cheaper, the use of the machine should be contained within the limit in which its own production requires less work than what its use replaces. For capital, however, the limit is tighter. Since it does not pay the labour employed, but the value of the labour force used, the application of machinery to capital is limited by the difference between the value of the machine and the value of the labour force replaced".[32]

---

[30] Das Kapital, p. 435.

[31] Das Kapital, p. 436.

[32] Das Kapital, p. 447.

The theoretical model developed by Marx to analyze the capitalist mode of production, which is the central objective of his greatest work, continues to be of fundamental importance in the present day, despite the profound transformations which capitalism has gone through during the almost 140 years since the publication of *Das Kapital.* There is a very simple reason for this. The material base on which capitalism rests preserves the same contradictions as the time of Marx, as well as having several others. The regime of the exploitation and appropriation of human labour through the surplus mechanism, the system of individual appropriation (private ownership of the means of production) despite socialized production, as well as its immense destructive capacity in relation to nature and life on the planet.

Attempts to update Marx`s ideas, principally those contained in *Das Kapital,* have been made by those who believe that his theoretical model is based on the existence of many different capitals participating in a hypothetical competitive market into which they flow, and it is this reason that Marx stated that the bourgeoisie have to constantly revolutionize the means of production.[33] In so-called modern capitalism, capital was concentrated, the trusts and cartels appeared and there was a fusion of banking capital with industrial giving rise to financial capital.[34] In the recent globalized capitalism what is most striking is the excessive predominance of financial capital, so much so that some economists speak about a new form of financial accumulation.[35] Nevertheless, the old question which pursued Marx continues to hover over our heads, no longer the specter which haunted Europe, but rather a need for the transformation of present day society on a global scale.

[33] Cláudio Napoleoni, op. cit. p. 141, states: "This means that the multiplicity of capital, and thus competition, constitute in Marx's theory an essential trait of capitalist reality."

[34] V. I. Lênin in "Imperialismo Etapa Superior do Capitalismo", Editora Centelha, Coimbra, 1974, p. 24/25 states: "Half a century ago when Marx wrote *Das Kapital,* free competition was seen by the immense majority of economists as a 'law of nature.' Official science tried to kill Marx's work in a conspiracy of silence, which demonstrated with a theoretical and historical analysis of capitalism, that free competition engendered the concentration of production, which, reaching a certain level of development, led to monopoly. Currently monopoly has become a fact." Referring to merger processes, Lenin notes the following on page 53: "At the same time there develops, it can be said, the personal union of banks and large industrial and commercial companies, the mergers of some with others, through the purchase of shares, the entrance of banks directors to the boards (or administrations) of industrial and commercial companies and vice-versa" (Lênin 1974).

[35] In his book *A Mundialização do Capital,* Xamã Editora, 1ª. Edição, 1996, p. 246, François Chesnais, states "The author who, based on the facts observed between 1860 and 1870, most clearly perceived the capacity of concentrated monetary capital living at the expense of the sphere of wealth creation was Marx. He shows the formation of an organized and concentrated mass of capital-money, which to the contrary of real production is placed under the control of bankers (Book III, Cap. 25). This mass allows that this capital stops being the simple link between the valorization of capital in industrial production, to constitute an independent force and the nest for the accumulation of financial profits" (1996).

References

Ampère AM (1823) Théorie Mathematique des Phénomenes Electro-Dynamiques Uniquement Déduite de l' Éxperience, éditions Jacques Gabay, Paris

Ashton TS (1995) Revolução Industrial, Publicações Europa-América, 2ª. Edição, Lisboa

Babbage C (1880) Economie des Machines et des Manufactures. Librairie du Dictionaire des Arts et Manufactures, Paris

Bachelard GA (1996) Formação do Espírito Científico, Editora Contraponto, 4a. reimpressão, Rio de Janeiro

Chenais FA (1996) Mundialização do Capital, Xamã Editora, 1ª. Edição

Crowe MJ (1993) A history of vector analysis. Dover Publications, Inc., New York

de Oliveira BJ (2002) Francis Bacon e a Fundamentação da Ciência como Tecnologia. B. Horizonte, Editora da UFMG

Ellul J (1968) A Técnica e o Desafio do Século. Editora Civilização Brasileira, Rio de Janeiro

Ellul J (2004) Le Système Technicien. La Cherche Midi, Paris

Galilei G (1988) Discurso sobre as Duas Novas Ciências. Museu de Astronomia e Ciências Afins, Rio de Janeiro

Grange J La (1996) Philosophie d'Auguste Comte. Presses Universitaires de France, Paris

Horizonte B (2005) Proceedings of seminar on history of science and technology (in Portuguese)

Jung J (2000) Le travail. Flammarion Éditeurs, Paris

Koyré A (1963) Galileu e Platão, Editora Gradiva, Portugal

Leff E (2000) Complexidade, Interdisciplinaridade e Saber Ambiental, paper published in *Interdisciplinaridade em Ciências Ambientais*, p 19, Signus Editora, S. Paulo

Lênin VI (1974) Imperialismo Etapa Superior do Capitalismo. Editora Centelha, Coimbra

Mantoux PA (1988) Revolução Industrial no Século XVIII, Editora UNESP

Napoleoni C, Smith R (1988) Marx, Editora Graal, 6a. Edição

Pedretti C (1999) Leonardo: The machines. Giunti Editors, Florence

Romero D (2005) Marx e a Técnica. Editora Expressão Popular, S. Paulo

Rossi P (1989) Os Filósofos e as Máquinas. Editora Schwarcz Ltda., S. Paulo

Simondon G (2005) L'Invention dans les Techniques. Éditions du Seuil, Paris

Thuillier P (1976) Au Commencement Etait La Machine. La Recherche, p. 47

Ure A (1836) Philosophie des Manufactures et Économie Industrièlle, Brussèle

Erratum to: The Conceptual Basis to Work Studies

Agamenon R. E. Oliveira

Erratum to: Chapter 2 in: A. R. E. Oliveira, *A History of the Work Concept*, History of Mechanism and Machine Science, DOI 10.1007/978-94-007-7705-7_2

Chapter 2 contains an unfortunate error in the footnotes.

The following footnote should be added as further information on Descartes's *Geometry* mentioned on page 49:

Descartes's *Geometry* is a book of 87 pages divided into three parts: Book I—Problems which we can build with circles and straight lines only; Book II—On the nature of curved lines; Book III—On the construction of problems which are solids or more than solids. It deals with the study of problems of geometry becoming algebraic equations. On page 3 of Book I Descartes describes his method: Thus, to solve any problem, we should use each in order, first considering those already done, and find values for all lines to be constructed, also for those unknown. Then, without considering any difference between the known and unknown lines, we should proceed to solve them in the order that presents itself the most naturally of all such that they depend mutually one on the other, until one can find a means to express an equal quantity of two ways, what we call equation; because the terms of one of these two manners are equal to the other. We should find as many equations as lines that are unknown (Descartes 1991).

The online version of the original chapter can be found under
DOI 10.1007/978-94-007-7705-7_2

A. R. E. Oliveira (✉)
Applied Mechanics and Structures, Polytechnica School of Rio de Janeiro, Federal University of Rio de Janeiro, Rio de Janeiro, Brazil
e-mail: agamenon.oliveira@globo.com

A. R. E. Oliveira, *A History of the Work Concept*, History of Mechanism and Machine Science 24, DOI: 10.1007/978-94-007-7705-7_9,

In addition, the reference to footnote 9 in the middle of page 50, actually refers to footnote 8 as it appears at the bottom of page 49. Footnote 10 in the main text refers to footnote 9 at the bottom of page 50, etc., to the end of the chapter.

Lastly, footnotes 22 and 21 should be switched around, so they appear in correct numerical order.

MIX
Papier aus verantwortungsvollen Quellen
Paper from responsible sources
FSC® C105338

If you have any concerns about our products,
you can contact us on
ProductSafety@springernature.com

In case Publisher is established outside the EU,
the EU authorized representative is:
Springer Nature Customer Service Center GmbH
Europaplatz 3, 69115 Heidelberg, Germany

Printed by Libri Plureos GmbH
in Hamburg, Germany